Contents

Introduction

The purpose of this revision book is to help you fulfil your potential when you take GCSE Chemistry or the Chemistry component of Double Award Science (Modular or Non-modular), at either Foundation or Higher tier. It is a companion to *GCSE Chemistry for CCEA*.

Chemistry at GCSE is divided into two types of material: *recall* and *application*. Recall is the factual information that you need to learn. Application is the use of this factual information in calculations or in writing formulae and equations. Recall makes up about two-thirds of the examination papers. Most of this book consists of 'recall material' – the key information that you need to know. For those of you aiming at a grade B or C, the recall material will give you good grounding in essentials. Parts of the book involving application of information are indicated by ➜. This 'application material' is more difficult, but for those aiming at a high grade it is extremely important.

In each chapter the most important points of each topic are explained, and understanding and practice at application are built up through the use of examples, typical questions and worked answers. Guidance in examination technique is given in the form of examiner's notes and hints. These highlight common misconceptions and mistakes made by students.

Take note of the stripes down the right-hand edge of some pages:

- parts of the book that are only applicable if you are working at Higher tier of either Double Award Science or GCSE Chemistry are identified by a **grey stripe**
- parts relevant only if you are taking GCSE Chemistry have a **black dotted stripe**
- parts relevant only if you are taking Higher GCSE Chemistry have a **black solid stripe**.

At the end of each chapter there are revision questions to test your knowledge. Answers to these revision questions are at the back of the book.

This book provides you with a carefully planned revision strategy, reinforcing core knowledge and essential understanding. It will enable you to achieve your very best, whether you are striving to obtain an A or A* grade or whether you are hoping to achieve a grade C.

1

States of matter

There are three states of matter:

solid **liquid** **gas**

Arrangement of particles

The arrangement of particles is different in each state and is shown in Figure 1.1.

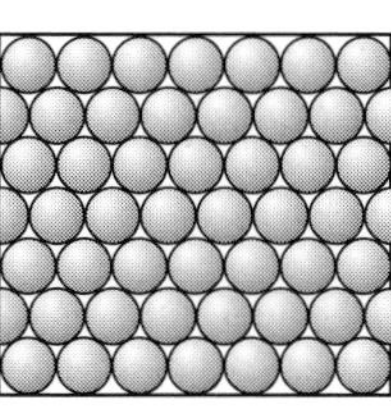

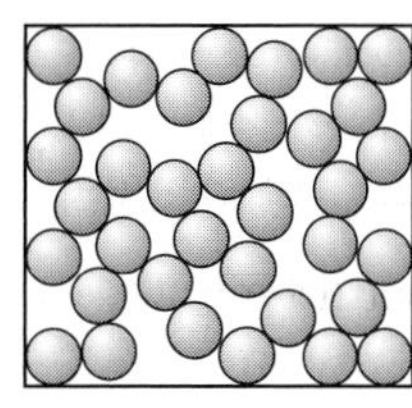

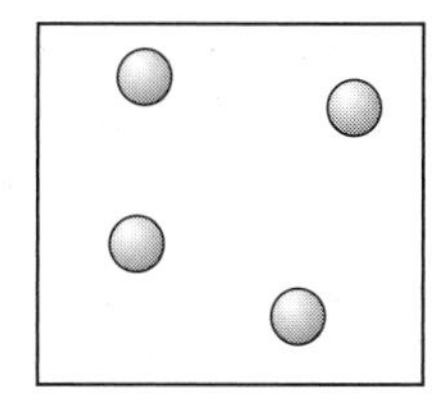

Figure 1.1 The arrangement of particles in a solid, a liquid and a gas

Physical properties

The **physical properties** of each state are shown in Table 1.1.

Table 1.1 Some general physical properties of the three states of matter

Physical property	Solid	Liquid	Gas
shape	fixed shape	takes the shape of sides and base of container	takes the shape of whole container
volume	fixed	fixed	not fixed – as large as container
flow	does not flow, rigid	flows	flows easily
compressibility	cannot be compressed	cannot be compressed	can be compressed
density	high density	medium density	low density
expansion on heating	low expansion	medium expansion	high expansion

NOTE: You will often be asked about the physical properties of solids, liquids and gases in a Recall question. Explaining these properties in terms of the particles is a common type of Application question.

The properties of the three states of matter can be explained using five different features – see Table 1.2.

NOTE: A good way of remembering these features is:

'EMAPA': Energy, **M**otion, **A**rrangement, **P**roximity and **A**ttraction

Table 1.2

Feature	Solid	Liquid	Gas
Energy/speed of particles	low energy/very little movement	medium energy/some movement	high energy/fast movement
Motion of particles	only vibrate	can move around each other	random movement of particles
Arrangement of particles	regular arrangement	no regular arrangement	no regular arrangement
Proximity of particles	closely packed together	closely packed together	well spaced out from each other
Attraction of particles	high forces of attraction	high forces of attraction	virtually no forces of attraction

NOTE: The characteristics of the particles in each of the states are needed to explain the physical properties and **changes of state** in Application questions.

Changes of state

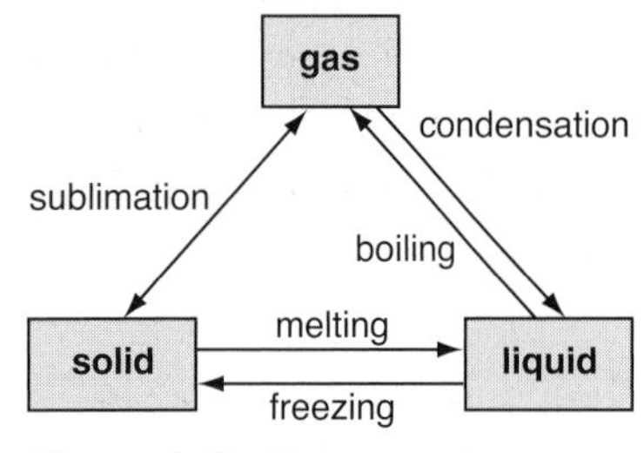

Figure 1.2 Changes of state

Figure 1.2 shows the changes of state.

The following are definitions that you need to know relating to the changes of state in Figure 1.2:

- **melting point** is the temperature at which a solid changes to a liquid
- **boiling point** is the temperature at which a liquid changes to a gas
- **evaporation** is the change of state from a liquid to a gas below the boiling point
- **sublimation** is the direct change from a solid to a gas (or from a gas to a solid).

Examples of substances that undergo **sublimation** when they are heated are carbon dioxide and iodine. Solid carbon dioxide is called dry ice. Iodine changes from a dark grey solid to a purple gas when it is heated.

HINT In your answer to a question that asks 'What is melting?', you should include *both* states. For example, 'melting is the change from a solid to a liquid'.

The only melting and boiling points you will be expected to remember are those of water – water melts at 0 °C and boils at 100 °C. Temperature is usually measured in units of degrees Celsius (°C). For certain calculations, temperature is measured in kelvin (K) – note there is no degree symbol in this unit.

NOTE: Melting and boiling points should be given in degrees Celsius (°C) when a question asks you to determine a state at room temperature and pressure. Only state temperature in kelvin (K) in a pressure, volume and temperature (*PVT*) type question.

Kinetic theory

The **kinetic theory** states that all matter is composed of small particles which are invisible even under a light microscope. Particles in different substances are different sizes.

The following pieces of evidence support the kinetic theory.

- **Brownian motion:** a sample of smoke shows random jittery specks of light under a microscope – caused by the air particles bombarding the solid smoke particles.
- **Diffusion:** movement of particles of one substance through particles of another substance so that the substances mix. Diffusion is fastest in a gas as the particles are far apart and slowest in a solid as the particles are close together.
- **White-ring experiment:** a glass tube has cotton wool soaked in concentrated ammmonia (NH_3) at one end and cotton wool soaked in concentrated hydrochloric acid (HCl) at the other. A white ring appears in the tube closer to the hydrochloric acid end. The HCl gas particles are heavier and larger than the NH_3 gas particles. The lighter NH_3 particles move further and faster.

NOTE: Diffusion questions are usually about gases, especially smelly gases or coloured gases, as these can be easily detected.

The movement of particles is faster at higher temperature so diffusion happens more quickly at higher temperatures.

Using kinetic theory to explain the three states

The kinetic theory explains the energy changes during changes of state.

- As a solid substance is heated the particles gain energy and vibrate more – this often makes solids expand on heating.
- When the particles in a solid get enough energy to break the attraction between them, the solid melts – this happens at its **melting point** (see Figure 1.3 on page 4).

- When the particles in a liquid get enough energy to escape from the surface of the liquid, the liquid boils and becomes a gas – this happens at its **boiling point**.
- In a liquid below its boiling point, some particles can escape from the surface of the liquid. This is called evaporation.

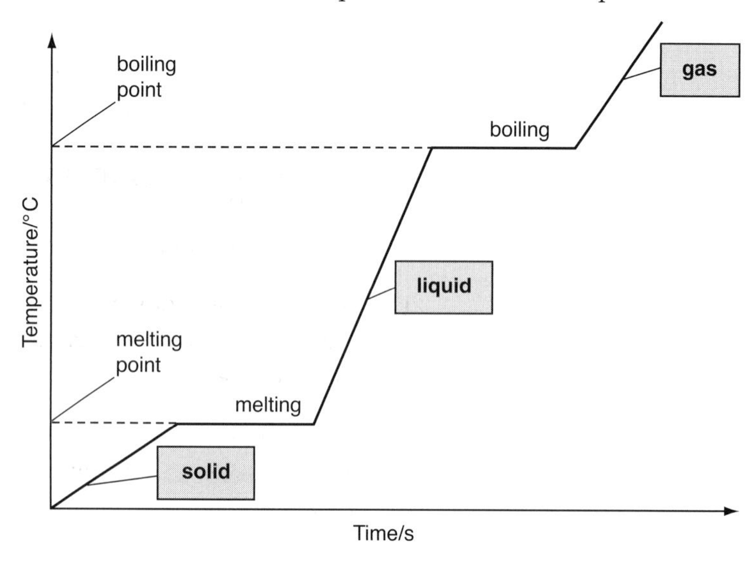

Figure 1.3 The graph shows the temperature during changes of state

Determining state from melting and boiling points

When the melting and boiling points of a substance are given, the state at any temperature can be determined as shown in the worked examples that follow.

Example 1

Bromine's melting point is −7 °C and its boiling point is 59 °C.

- At temperatures below −7 °C, bromine is a solid.
- At temperatures between −7 °C and 59 °C, bromine is a liquid.
- At temperatures above 59 °C, bromine is a gas.

On a temperature line, mark the melting point and boiling point, with the lowest or most negative value furthest to the left.

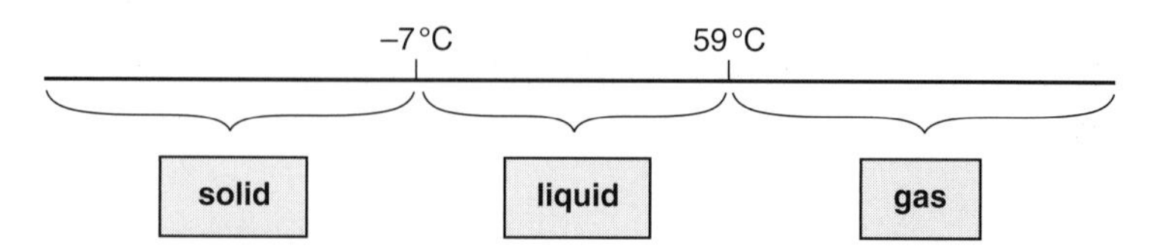

Figure 1.4

- At all temperatures *to the left* of (below) the melting point, the substance is a solid.
- At all temperatures *between* the melting and boiling points, the substance is a liquid.
- At all temperatures *to the right* of (above) the boiling point, the substance is a gas.

Example 2

Figure 1.5

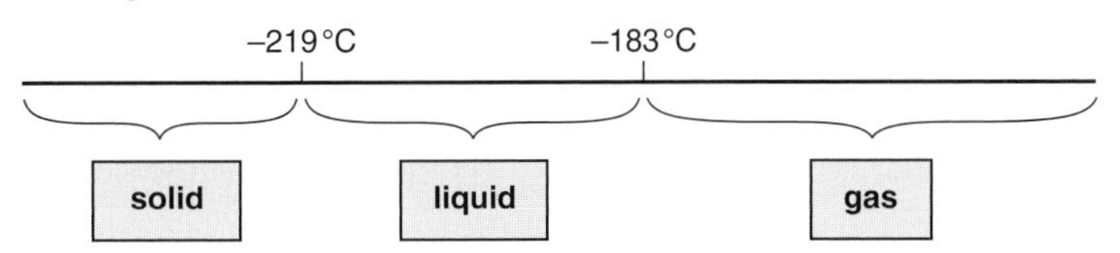

Oxygen's melting point is −219 °C and its boiling point is −183 °C.

- At −191 °C oxygen is a liquid.
- At 45 °C oxygen is a gas.
- At −230 °C oxygen is a solid.

Explanations using 'EMAPA'

All properties, observations and changes of state can be explained using 'EMAPA' (see page 2 to remind yourself of its meaning).

Example 3

Explain, in terms of particles, what happens when a solid melts.

1 Particles gain energy.
2 Particles start to move more.
3 Particles move away from their fixed positions.
4 Particles move further apart from each other.
5 Particles are less attracted to each other.

Typical questions

1 Explain what happens to the particles in ice when an ice cube melts. [4]
2 Explain what happens to the particles in a solid when it dissolves in water. [3]

Answers

1 *Particles gain energy.* [1]
Particles vibrate more. [1]
Attraction between the particles decreases. [1]
The particles can now move from their fixed position. [1]

NOTE: The most common mistake in this question is to state that 'the particles gain energy and begin to vibrate' – the particles are already vibrating but they vibrate more.

2 *Water particles move and hit solid particles.* [1]
The solid breaks up. [1]
Particles of solid move in between the water particles. [1]

NOTE: The most common mistake in this type of question is to miss the point completely and write about the solid being soluble – the question asks in terms of particles and your answer should relate to particles.

Pressure, volume and temperature (*PVT*) calculations

The **volume (*V*)** of a gas depends on its **temperature (*T*)** and **pressure (*P*)**.

- An increase in the temperature of a gas will increase its volume.
- A decrease in the temperature of a gas will decrease its volume.
- An increase in pressure will decrease the volume of a gas.
- A decrease in pressure will increase the volume of a gas.

Typical question

1 How does the volume of a gas change as pressure changes? *[2]*

Answer

1 *As the pressure of the gas increases,* *[1]*
its volume decreases. *[1]*

NOTE: The most common mistake here is to forget to state how pressure changes first and then how it affects the volume – a common mistake is just to state that 'the volume increases', which gets 0 marks as it makes no sense.

The general gas equation

$$\frac{P_1V_1}{T_1} = \frac{P_2V_2}{T_2}$$

where P_1V_1 means $P_1 \times V_1$, and P_2V_2 means $P_2 \times V_2$.

There are many units of pressure and volume that can be used in *PVT* calculations.

- Pressure units are usually atmospheres (atm), but can also be pascals (Pa), kilopascals (kPa), newtons per square metre (N/m^2).
- Volume units are usually cubic centimetres (cm^3), but can also be cubic metres (m^3), cubic decimetres (dm^3).
- Temperature units in *PVT* questions are *always* kelvin (K).

HINT: In a *PVT* question, 1 mark is usually awarded for writing the general gas equation.
Remember that temperature is always given in kelvin. In Double Award Science, the general gas equation is usually written $\frac{PV}{T} = k$, where '*k*' is a constant. This equation is often given in the question and it means the same as $\frac{P_1V_1}{T_1} = \frac{P_2V_2}{T_2}$.

Calculations using the gas equation

➜ All of P, V and T are given as one set of conditions – P_1, V_1 and T_1. Another set of conditions is given where one of the three (usually V) is missing. These are P_2, V_2 and T_2, and usually V_2 is the one you need to calculate.

Example

A sample of methane has a volume of 25 cm^3 at a temperature of 200 K and a pressure of 2 atm. Calculate the volume the same sample of gas will occupy at 400 K and 5 atm pressure.

1 Write out the general gas equation and all the values of PVT as a complete set of P_1, V_1, T_1 and a set of P_2, V_2 and T_2, with the unknown value – the one you need to find – missing.

$$\frac{P_1V_1}{T_1} = \frac{P_2V_2}{T_2}$$

$P_1 = 2$ atm $P_2 = 5$ atm
$V_1 = 25$ cm^3 $V_2 = ?$ (the one you need to find)
$T_1 = 200$ K $T_2 = 400$ K

2 Write the values into the general gas equation:

$$\frac{2 \times 25}{200} = \frac{5 \times V_2}{400}$$

3 Calculate the left-hand side of the equation:

$$\frac{2 \times 25}{200} = 0.25 \qquad \text{so: } \frac{5 \times V_2}{400} = 0.25$$

4 Now you can calculate V_2 by rearranging the equation:

$$V_2 = \frac{0.25 \times 400}{5} = 20$$

$V_2 = 20$ cm^3

NOTE: Always give your answer in the correct units – in this example they are the same as the units for the value V_1.

NOTE: Questions can also be set to find a pressure, and they follow the same pattern as in the example above. If one of P, V or T is not given in the question, you can assume it did not change (or the question may tell you this) and you can make it equal to 1.

Typical question

2 A gas has a volume of 125 dm^3 at 200 K. What would the volume be at 325 K, assuming the pressure remained constant? *[4]*

Answer

2 $\frac{P_1V_1}{T_1} = \frac{P_2V_2}{T_2}$ $P_1 = 1$ atm $P_2 = 1$ atm
$V_1 = 125$ dm^3 $V_2 = ?$ *[1]*
$T_1 = 200$ K $T_2 = 325$ K

$$\frac{1 \times 125}{200} = \frac{1 \times V_2}{325}$$ *[1]*

$$0.625 = \frac{V_2}{325}$$

$$V_2 = 0.625 \times 325 = 203.125\ dm^3$$ *[2]*

Revision questions

1 How does the volume of a gas change as temperature changes? *[2]*

2 How are the particles arranged in a solid? *[2]*

3 Explain why gases are compressible. *[2]*

4 State **two** physical properties of liquids. *[2]*

5 A gas has a volume of 150 cm^3 at 200 K and 2.5 atm pressure.
Calculate the volume of the gas at 150 K and 1.5 atm. *[4]*

6 What is the difference between melting and sublimation? *[2]*

7 Name **one** substance that undergoes sublimation when heated. *[1]*

8 A sample of nitrogen is compressed to 2 atm at 200 K and has a volume of 25 dm^3.
What volume is obtained when the temperature is increased to 300 K at the same pressure? *[4]*

9 What is the name of the change of state from gas to liquid? *[1]*

10 Concentrated ammonia solution and concentrated hydrochloric acid are placed at opposite ends of a glass tube. A white rings forms in the tube.

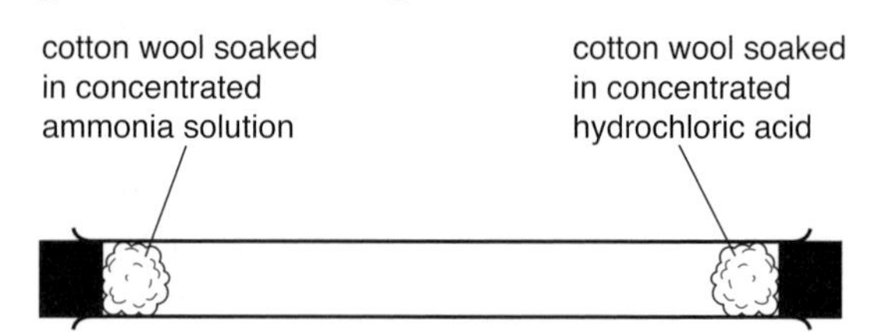

a Make a sketch of the diagram and mark on it approximately where in the tube the white ring forms. *[1]*
b Explain why the white ring forms in this position. *[2]*

11 What name is given to the random movement observed when smoke is viewed through a microscope? *[1]*

12 Explain, in terms of particles, what happens when a sample of sugar dissolves in water. *[3]*

13 Mercury melts at −39 °C and boils at 630 °C. In what physical state would you find mercury at:
a −200 °C **b** 20 °C **c** 1000 °C? *[3]*

14 What is the common name for solid carbon dioxide? *[1]*

15 What is observed when solid iodine is heated? *[4]*

2

The Periodic Table

You need to know the following definitions.

- The **Periodic Table** lists all known elements.
- An **element** is a substance that is made up of only one type of atom.
- A **compound** is a substance that consists of two or more elements that are chemically combined.
- An **atom** is the simplest particle of an element which can exist on its own in a stable environment.
- A **molecule** is a particle that consists of two or more atoms chemically bonded together.

History of the Periodic Table

- 1864 – John Newlands arranged the elements in order of atomic mass and found the first element was similar to the eighth, and the second was similar to the ninth. He called this pattern the **law of octaves** (the Noble Gases had not yet been discovered).
- 1869 – Dimitri Mendeleev also arranged the elements in order of atomic mass, but left gaps for undiscovered elements and switched the mass order to fit the patterns in the table (the Noble gases had still not been discovered!)
- The modern Periodic Table we use today is arranged in order of atomic number.

Typical question

Describe the work of Mendeleev in the development of the modern Periodic Table. *[3]*

Answer

Arranged elements in order of atomic mass [1]. Left gaps [1] for undiscovered elements [1]. Changed the order of certain elements to suit properties [1].

NOTE: This type of question is marked as 3 out of a possible 4 marks. You only need 3 of the 4 points in the answer to gain the full 3 marks.
The most common mistake in this question is to state that Mendeleev arranged the table in order of atomic number. Remember that Mendeleev also changed the order of some of the elements, such as iodine and tellurium, to better suit their properties.

Groups and periods

- The **periods** are the horizontal rows of the Periodic Table.
- The **groups** are the vertical columns of the Periodic Table.
- The first period contains only two elements – hydrogen and helium.

Figure 2.1 The modern-day Periodic Table, showing the positions of the periods and groups

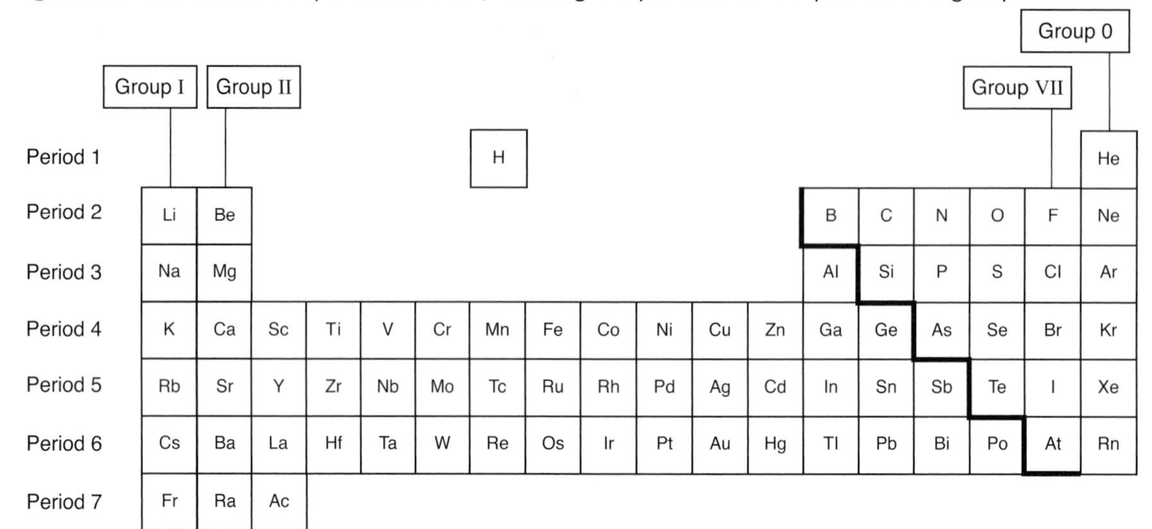

HINT: A common question is to name or state the symbol of an element in a certain period and a certain group of the Periodic Table. Remember to count Period 1. For example, the element in Period 3 and Group VII is chlorine. A common mistake is to write 'bromine', having forgotten to count Period 1.

The thick black line shown on the Periodic Table above divides metals (on the left of the line) from non-metals (on the right). It is not shown on the Periodic Table in the *Data Leaflet* in the examination. It is a good idea to draw it onto the Periodic Table when you get it in the exam. Remember to go down the left-hand side of boron and continue in steps as shown.

Metals, non-metals and semi-metals

- The elements to the left of the thick black line in the Periodic Table are **metals**.
- The elements to the right of the thick black line in the Periodic Table are **non-metals**.
- Two elements beside the black line (silicon, Si and germanium, Ge) are described as **semi-metals**.
- A semi-metal is an element that exhibits properties of both metals and non-metals.

NOTE: The only two semi-metals that are accepted as answers at GCSE are silicon and germanium.

Group names

- **Group I** is called the **alkali metals**.
- **Group II** is called the **alkaline earth metals**.
- **Group VII** is called the **halogens**.
- **Group 0** (also called **Group VIII**) is called the **Noble gases**.

HINT: The names of Group I and Group II are often confused. Common errors are to call Group I the alkal*ine* metals and Group II the *alkali* earth metals.

Solids, liquids and gases

- Of all the known elements:
  - 11 are gases
  - 2 are liquids
  - the rest are solids.
- The 11 gases are hydrogen, nitrogen, oxygen, fluorine, chlorine and the Noble gases (helium, neon, argon, krypton, xenon, radon).
- The 2 liquids are bromine (a **non-metal**) and mercury (a **metal**).

Diatomicity

Some elements exist as **diatomic** molecules.

- Diatomic means that two of the same atoms are joined together by a covalent bond.
- When writing the formulae of these elements in a balanced symbol equation they should be written with a small '2' after the symbol to indicate two atoms joined as a molecule, for example H_2, O_2.
- There are seven diatomic elements to remember:
  - hydrogen, H_2
  - nitrogen, N_2
  - oxygen, O_2
  - fluorine, F_2
  - chlorine, Cl_2
  - bromine, Br_2
  - iodine, I_2.

NOTE: If you are asked for the symbol for an element or an atom of the element, it is correct to write Cl or H, for example, rather than Cl_2 or H_2.

Properties of the elements

Similarities between elements in Group I

- All are very reactive metals.
- All are soft, easily cut metals that expose a shiny surface when freshly cut.
- The shiny surface quickly tarnishes (goes dull) in air.
- All have low melting points.
- All conduct electricity.
- All react vigorously with water and burn in air.
- All atoms of Group I elements have 1 electron in their outer shell.
- All form simple ions with a charge of '+', for example Na^+, K^+.
- All have a valency of 1.

Typical question

In terms of electronic configuration, what do the elements of Group I have in common? *[2]*

Answer

One electron [1] in their outer shell [1].

NOTE: The most common mistake here is to miss the last mark, for which you need to state 'in their outer shell'.

Similarities between elements in Group II

- All are reactive metals.
- All conduct electricity.
- All react with dilute hydrochloric acid and dilute sulphuric acid to produce hydrogen gas.
- Magnesium reacts very slowly with water but rapidly with steam to produce magnesium oxide and hydrogen gas.
- Calcium, strontium and barium react rapidly with water to produce hydrogen gas.
- All atoms of Group II elements have 2 electrons in their outer shell.
- All form simple ions with a charge of '2+', for example Mg^{2+}, Ca^{2+}.
- All have a valency of 2.

Similarities between elements in Group VII

- All are reactive non-metals.
- All are coloured.
- All exist as diatomic molecules.
- They do not conduct electricity.
- All react with Group I elements to form solid white ionic compounds.
- All atoms of Group VII elements have 7 electrons in their outer shell.
- All form simple ions with a charge of '−', for example Cl^-, Br^-.
- All have a valency of 1.

Similarities between elements in Group 0

- All are unreactive, colourless, non-metal gases.
- All atoms of Group 0 elements have 8 electrons in their outer shell.
- Group 0 elements do not have a valency as they are unreactive.

Trends going down all groups

- Atomic size increases: the outermost electron is further from the nucleus as a group is descended, making the atom bigger.
- Relative atomic mass increases: the number of protons (and neutrons) increases as a group is descended, which increases the relative atomic mass.

Specific trends going down Groups I and II (metals)

- Reactivity increases.
- Melting point decreases.

Specific trends going down Group VII

- Reactivity decreases.
- Colour intensity of the elements darkens: F_2 is a pale yellow gas; Cl_2 is a yellow-green gas; Br_2 is a red-brown liquid; I_2 is a dark grey solid. (Iodine gas is purple and the dark grey solid iodine sublimes to form the purple gas when heated.)
- Based on trends, it would be expected that astatine is a black solid.
- Melting point increases.

Specific trend going down Group 0

- Density increases (helium and neon are less dense than air and the others are more dense than air).

Trends going across Periods 2 and 3

- Atomic size decreases.
- Character changes from being metal to semi-metal to non-metal.
- Nature of the oxide changes from basic to amphoteric to acidic.
  - Metal oxides such as sodium oxide and magnesium oxide are described as **basic oxides**.
  - A basic oxide is one that reacts with an acid.
  - Non-metal oxides such as sulphur dioxide, carbon dioxide and nitrogen dioxide are described as **acidic oxides**.
  - An acidic oxide is one that reacts with an alkali.
  - Some metal oxides react with both acids and alkalis and these oxides are called **amphoteric oxides** (aluminium oxide and zinc oxide are the ones you need to know).
- Valency increases from 1 to 4 (at Group IV) and decreases to 1 again (at Group VII).

HINTS:

- The difference in the trend in reactivity of the Group I and II elements compared to the Group VII elements is often confused. Questions are set which ask for the most reactive element in Group VII. Fluorine is the correct answer. Many students incorrectly state astatine or iodine.
- Sometimes an incomplete form of the Periodic Table is given containing only a few of the elements and in the questions that follow you are asked to use *only* the elements given. Many students include other elements in their answers that cannot be accepted.
- Stating the wrong trend in atomic size going across a period is a common mistake. Make sure you remember that atoms get *smaller* going across a period.
- Learn the colour and state of the halogens. You will need to know these for questions in this section or in non-metals.

Applying trends to unfamiliar elements

An unfamiliar element is usually one further down the Periodic Table. Examples of elements commonly used in questions and which you may not feel so confident about are rubidium, caesium, strontium, barium and astatine. Apply the trends you have learned to tell you how these elements would appear or react.

Limitations of classification

- The Periodic Table classifies all elements, but it is not a clear-cut system at times.
- Elements are separated into metals and non-metals, but some (such as silicon and germanium) do not fit easily into either group and so are called semi-metals.
- Non-metals do not conduct electricity, whereas metals do. Carbon in its graphite form also conducts electricity, even though it is a non-metal.
- All metals are solids at room temperature and pressure, except mercury which is a liquid.

Revision questions

1 What are the common names given to:
a Group I **b** Group II
c Group VII **d** Group 0? *[4]*

2 Name **one** element in Period 3 that forms a basic oxide. *[1]*

3 Which element is in Period 3 and Group V? *[1]*

4 Which scientist devised the law of octaves? *[1]*

5 Give the state and colour at room temperature and pressure of the following elements.
a fluorine **b** chlorine **c** bromine
d iodine **e** neon *[5]*

6 Write the formula of the following simple ions.
a rubidium **b** strontium
c lithium **d** fluoride *[4]*

7 Silicon is described as a semi-metal. Name **one** other semi-metal and explain what is meant by the term semi-metal. *[2]*

8 A piece of freshly cut sodium is shiny but tarnishes in air. What does 'tarnish' mean? *[1]*

9 State the trend observed in atomic size across Period 3. *[1]*

10 Using only the list of elements below answer the following questions.

hydrogen **sodium** **magnesium**
oxygen **calcium** **aluminium**
silicon **carbon** **nitrogen** **sulphur**

a Which **three** elements are diatomic? *[3]*
b Name **one** solid non-metal. *[1]*
c Name **one** element that forms an amphoteric oxide. *[1]*
d Name **one** element that forms an acidic oxide and give the formula of this oxide. *[1]*
e Name **one** element that reacts with dilute hydrochloric acid but does not react with water. *[1]*

11 Explain what you understand by 'basic oxide'. *[1]*

12 What is used to order the elements in the modern Periodic Table? *[1]*

13 What does amphoteric mean? *[1]*

14 Describe how valency changes across Period 3. *[2]*

15 Which is the most reactive halogen? *[1]*

3

Atomic structure and bonding

- An **atom** is the simplest particle of an **element** which can exist on its own in a stable environment.
- Atoms are made up of three subatomic particles: **protons**, **neutrons** and **electrons**.
- Protons and neutrons are found in the **nucleus** (the centre of the atom) and electrons are found in **shells** orbiting the nucleus.
- The mass of an atom is largely in the nucleus as the mass of an electron is very small compared to the mass of a proton or a neutron.

Atoms are **electrically neutral** (they have no charge). This is because they have equal numbers of protons and electrons. Protons have a positive charge and electrons have a negative charge. Neutrons do not have a charge.

Table 3.1 gives the relative masses and relative charges of the three subatomic particles.

Table 3.1 Relative masses and charges of a proton, neutron and electron and their position in the atom

Subatomic particle	Relative mass	Relative charge	Location within the atom
proton	1	+1	nucleus
neutron	1	0	nucleus
electron	$\frac{1}{1840}$	−1	shells

NOTE: An incomplete form of this table is often given in a question and you are asked to complete it. The most common mistakes are leaving the '+' out of '+1' in the relative charge of a proton; giving '0' as the relative mass of an electron; and giving the relative charge of a proton as '+' (instead of '+1') and of an electron as '−' (instead of '−1'). Learn Table 3.1!

Atomic number and mass number

- The **atomic number** is used to order the elements in the Periodic Table.
- The atomic number is also called the **proton number** as it tells us the number of protons in the nucleus of an atom.

- The **mass number** for a particular atom is the total number of protons and neutrons in its nucleus. (The electron mass is so small that it does not affect the mass number.)
- The number of electrons equals the number of protons in an atom. (This is not always the case but it is true *in an atom* as atoms are electrically neutral.)

The atomic number and mass number of an atom are usually written before the symbol of the element. The mass number is written at the top and the atomic number at the bottom, for example ${}^{12}_{6}\mathrm{C}$, ${}^{39}_{19}\mathrm{K}$, ${}^{35}_{17}\mathrm{Cl}$, ${}^{31}_{15}\mathrm{P}$, ${}^{27}_{13}\mathrm{Al}$.

Determining the number of subatomic particles

The number of subatomic particles in an atom can be determined from the atomic number and the mass number.

Number of protons = atomic number
Number of neutrons = mass number − atomic number
Number of electrons = number of protons

Example

Determine the numbers of subatomic particles in each of the following atoms.

${}^{12}_{6}\mathrm{C}$ atomic number = 6 (**6 protons**)

6 electrons as it is an atom

mass number − atomic number = 12 − 6 = 6 (**6 neutrons**)

${}^{39}_{19}\mathrm{K}$ atomic number = 19 (**19 protons**)

19 electrons as it is an atom

mass number − atomic number = 39 − 19 = 20 (**20 neutrons**)

${}^{35}_{17}\mathrm{Cl}$ atomic number = 17 (**17 protons**)

17 electrons as it is an atom

mass number − atomic number = 35 − 17 = 18 (**18 neutrons**)

HINT: Remember that atoms have the *same number* of protons and electrons. They do not have the same number of protons and neutrons (${}^{12}_{6}\mathrm{C}$ does, but it is not always the case).

Isotopes and relative atomic mass

There are two types of chlorine atom that have different mass numbers: $^{35}_{17}Cl$, $^{37}_{17}Cl$. There are three types of hydrogen atom that have different mass numbers: $^{1}_{1}H$, $^{2}_{1}H$, $^{3}_{1}H$.

Atoms of the same element that have different mass numbers are called isotopes.

Isotopes are atoms with the same number of protons but different numbers of neutrons.

The two isotopes of chlorine, $^{35}_{17}Cl$ and $^{37}_{17}Cl$ (often called chlorine-35 and chlorine-37) are both chlorine atoms and have all the properties of chlorine atoms.

The term 'mass number' applies to a particular atom. The mass of a chlorine atom in a sample of chlorine is an average of the masses of chlorine-35 and chlorine-37. But only $\frac{1}{4}$ of chlorine atoms are chlorine-37 and $\frac{3}{4}$ are chlorine-35, giving the average mass of a chlorine atom to be 35.5. This average mass is called the **relative atomic mass**. The word 'relative' is used as the mass of all atoms is measured relative to the mass of an atom of carbon-12.

HINT: Being asked to define the term 'isotope' is a common question. Remember that there are three main parts to the definition: 'isotopes are *atoms* with the *same atomic number* (or number of protons) but with *different mass numbers* (numbers of neutrons)'.

History of the development of atomic theory

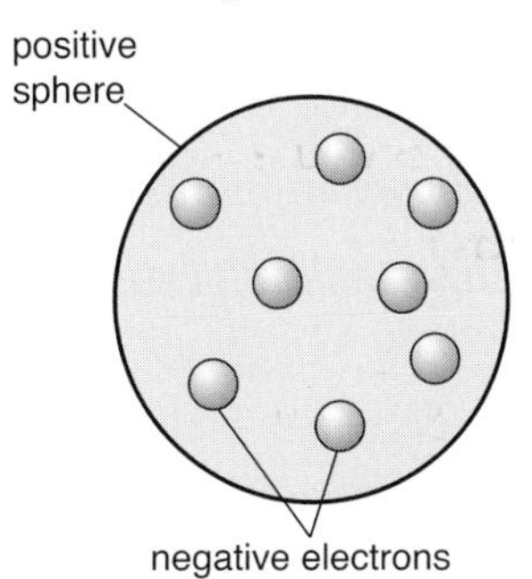

Figure 3.1 Thomson's 'plum pudding' atomic model

- Early-1800s – John Dalton stated that atoms were the smallest particles of matter.
- 1897 – JJ Thomson proposed the 'plum pudding' model which stated that the negative electrons were embedded in a positive sphere (like the raisins in a plum pudding – see Figure 3.1).
- 1911 – Ernest Rutherford revised JJ Thomson's model to the model which has electrons orbiting a positive nucleus.
- 1932 – James Chadwick discovered the neutron.

Electronic configuration

The arrangement of the electrons in the shells around the nucleus of an atom can be represented either as a diagram or in written format but remember the following rules (overleaf) as to where the electrons can be.

1. The shells are at an increasing distance from the central nucleus. The shell closest to the nucleus is called the first shell, then there are the second, third and fourth shells at increasing distances from the nucleus.
2. The first shell can hold a maximum of 2 electrons.
3. The second and third shells can hold a maximum of 8 electrons.
4. The first shell must fill first, before an electron can be put in the second shell and the second shell must fill before an electron can be put in the third shell, and so on.
5. Electrons pair up in a shell but only when no other space is available. For example, 4 electrons in shell 2 are not paired, but 6 electrons would have 2 pairs and 2 unpaired. The 2 electrons in the first shell must be paired as it can only hold a maximum of 2.

Example 1

Lithium has atomic number 3, so an atom of lithium has 3 electrons. The first shell takes 2 electrons and 1 electron goes into the second shell.

The written electronic configuration of lithium is: 2, 1.

The drawn electronic configuration is shown in Figure 3.2. Each cross represents an electron.

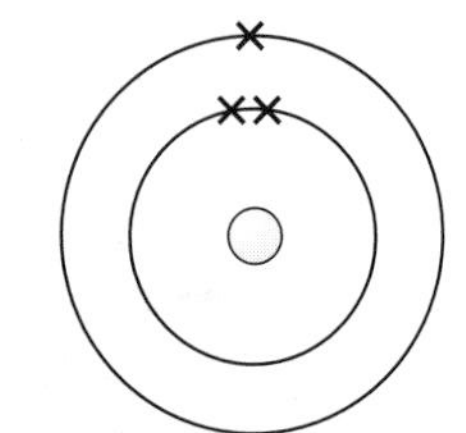

Figure 3.2 Electronic configuration of lithium

Example 2

Sulphur has atomic number 16, so an atom of sulphur has 16 electrons. The first shell takes 2 electrons, the second shell takes 8 electrons and the third shell takes 6 electrons.

The written electronic configuration of sulphur is: 2, 8, 6.

The drawn electronic configuration is shown in Figure 3.3.

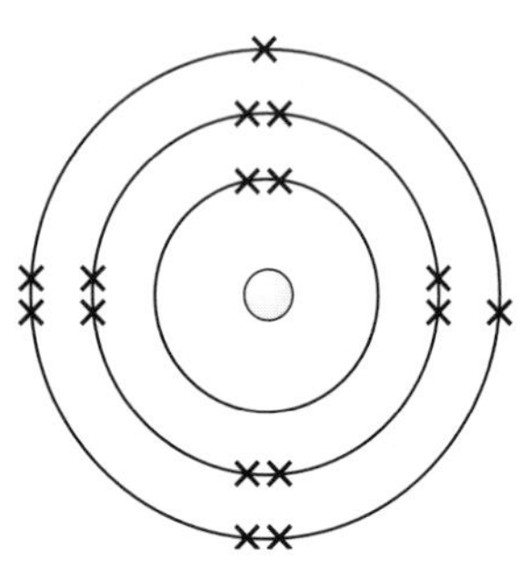

Figure 3.3 Electronic configuration of sulphur

The written and drawn electronic configurations for atoms of elements 1 to 20 are shown in Table 3.2.

Table 3.2 Electronic configurations for elements 1 to 20 (**x** = electron)

Hydrogen	Helium	Lithium	Beryllium	Boron
1	2	2, 1	2, 2	2, 3
Carbon	**Nitrogen**	**Oxygen**	**Fluorine**	**Neon**
2, 4	2, 5	2, 6	2, 7	2, 8
Sodium	**Magnesium**	**Aluminium**	**Silicon**	**Phosphorus**
2, 8, 1	2, 8, 2	2, 8, 3	2, 8, 4	2, 8, 5
Sulphur	**Chlorine**	**Argon**	**Potassium**	**Calcium**
2, 8, 6	2, 8, 7	2, 8, 8	2, 8, 8, 1	2, 8, 8, 2

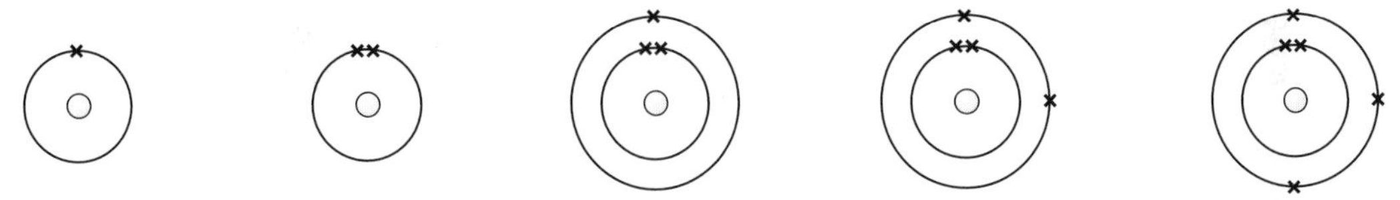

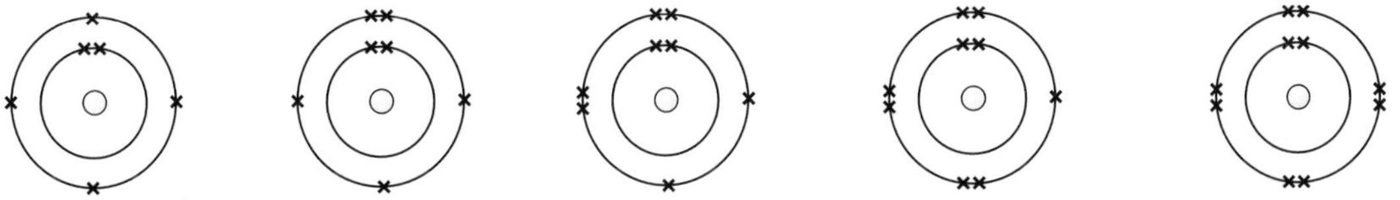

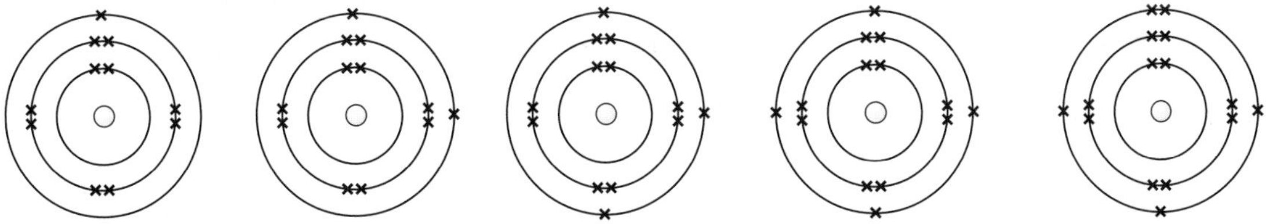

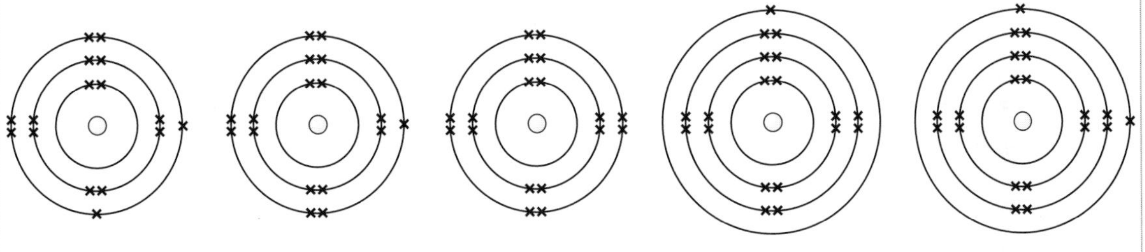

HINT: When a question asks you to draw a diagram of an atom you must write the correct number of protons and neutrons in the nucleus and the correct number and arrangement of electrons in the shells. These questions can be worth 4 marks.

If you are asked to show an electronic configuration, read the question twice to be sure if the question has asked for it in diagram or written form. If the question says 'draw', then make sure you do! If the question does not specify which, then the written form is acceptable.

Using the Periodic Table to determine electron configuration

The electronic configuration of an atom may be determined from the Periodic Table by using the group number and the period number.

group number = number of electrons in the outer shell
period number = number of shells in use

Example 3

What is the electronic configuration of potassium?
Potassium is in Period 4 and Group I, so:

Potassium	
Period 4	**–, –, –, –**
Group I	**–, –, –, 1**
Configuration	**2, 8, 8, 1**

Period 4 means 4 shells in use.
So the first three shells must be full (look at rule 4 on page 20).
Group I means 1 electron in the outer shell.
Electronic configuration (E.C.) of potassium is: 2, 8, 8, 1.

Example 4

What is the electronic configuration of nitrogen?
Nitrogen is in Period 2 and Group V, so:

Nitrogen	
Period 2	**–, –**
Group V	**–, 5**
Configuration	**2, 5**

Period 2 means 2 shells in use (first shell is full).
Group V means 5 electrons in the outer shell.
Electronic configuration (E.C.) of nitrogen is: 2, 5.

Noble gases

The Noble gases all have full outer shells. The electronic configurations for the Noble gases are:

helium, He: 2   neon, Ne: 2, 8   argon, Ar: 2, 8, 8

A full outer shell is stable and makes the Noble gases unreactive.

Bonding

- The way in which molecules and structures are held together is called **bonding**.
- There are three main types of bonding – **ionic**, **covalent** and **metallic**:
  - ionic bonding occurs in compounds containing a metal and a non-metal such as sodium chloride and magnesium oxide
  - covalent bonding occurs between non-metal atoms. This can be in compounds such as hydrogen chloride and water or in elements such as chlorine (remember Cl_2) and carbon (graphite)
  - metallic bonding occurs in metals.
- All bonding involves electrons and these are usually represented by dots (•) or crosses (×) in bonding diagrams. All electrons are the same. Using dots and crosses shows which atoms the electrons come from.

Structure

It is important to understand the structure of a substance as well as its bonding.

- Bonding and structure are *not* the same – a substance can be covalently bonded but it may have a simple structure or a giant structure.
- Structure and bonding are used to explain many of the properties of these substances.

Ionic bonding

Ionic compounds are compounds that contain a metal. They are said to have ionic bonding. Examples are: sodium chloride (NaCl), magnesium oxide (MgO) and calcium chloride ($CaCl_2$). Ionic compounds are made up of ions which are **charged particles**. An ionic compound contains a **positive ion** and a **negative ion**.

Formation of ions from atoms

When ionic compounds form from the atoms of its elements, a **transfer of electrons** occurs. Metal atoms lose electrons and give them to non-metal atoms, which accept electrons. Each will lose or gain enough electrons to give it a full outer shell and make it more stable. When a metal atom loses electrons, it becomes a positively charged ion. When a non-metal atom gains electrons, it becomes a negatively charged ion.

Example 1: Sodium chloride

A sodium atom has an electronic configuration of: 2, 8, 1.

When it reacts with a chlorine atom (configuration 2, 8, 7), the 1 outer electron of the sodium atom is given to the outer shell of the chlorine atom.

Sodium now has only 10 electrons (2, 8) but it has 11 protons in its nucleus so it has a charge of '+'.

The sodium atom is written 'Na'. The sodium ion formed is written 'Na^+'.

Simple positive ions have the same name as the atom, so it is called a **sodium ion**.

Chlorine now has 18 electrons (2, 8, 8) but it has 17 protons in its nucleus so it has a charge of '−'.

The chlorine atom is written 'Cl'. The chloride ion formed is written 'Cl^-'.

Simple negative ions change the end of their name to '**-ide**', so it is called a **chloride ion**.

The sodium and chloride ions are attracted to each other and form an **ionic compound**.

An ionic bond is the attraction between oppositely charged ions in an ionic compound.

Figure 3.4 summarises the process of forming an ionic bond in sodium chloride.

Figure 3.4 Ionic bonding in sodium chloride

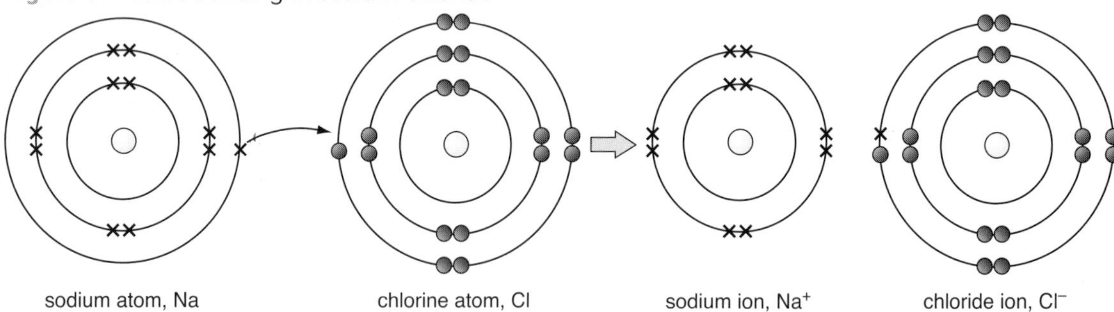

The compound formed is called sodium chloride (as it contains sodium ions and chloride ions). There is a lot of information here which helps in finding the **formula** of the compound. As a sodium atom loses 1 electron and a chlorine atom gains 1 electron, only one ion of each is required, so the formula of sodium chloride is **NaCl**. As a sodium ion is Na^+ and a chloride ion is Cl^-, only one of each ion is required so that the compound has no overall charge.

HINTS:

- Remember that compounds *do not* have a charge but ions *do* have a charge.
- When showing the formation of an ionic compound, always show the transfer of electrons with an arrow to make it clear where they are coming from and going to. You should also show the electronic configuration of the atoms and the ions. Make sure the charge on the ions is clear and the correct number of each ion is present.
- When an ion has a single positive charge, it is correct to write this as '+'. Do not write '1+' or '+1'. Similarly, when an ion has a single negative charge, write this as '−', not '1−' or '−1'.
- For ions with a higher charge, always write them in the same way as those on the back of the *Data Leaflet*, for example '2+' (*not* '+2'), '3−' (*not* '−3').
- Questions often ask about the electronic configuration of atoms and of ions. The electronic configuration of the Na^{+} ion is 2, 8 (see above) but this is the same as the electronic configuration of a Ne atom. The only difference is the number of protons in the nucleus. The number of electrons can help identify an unknown *atom* but be careful if it is an *ion*, as it will have lost or gained electrons. You can work out how many it has lost or gained from its charge.

Here are more examples of the bonding in ionic compounds.

Example 2: Magnesium oxide

Only one magnesium atom is required for each oxygen atom as magnesium atoms lose 2 electrons and oxygen atoms gain 2 electrons. A magnesium ion is Mg^{2+} as it has 10 electrons (2, 8) but has 12 protons (atomic number = 12).

An oxide ion is O^{2-} as it has 10 electrons (2, 8) but has 8 protons (atomic number = 8).

The formula of magnesium oxide is MgO.

Figure 3.5

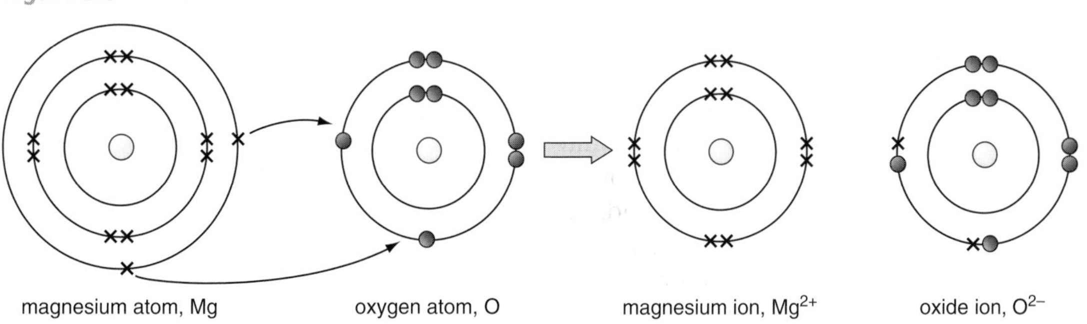

The charge on a simple ion can be checked by looking at the group number of the element. Group I elements form ions with a '+' charge, Group II '2+', Group III '3+' and Group IV does not generally form ions. Group V forms ions with a '3−' charge, Group VI '2−' and Group VII form ions with a '−' charge. Group 0 does not form ions as the atoms have full outer shells.

Example 3: Calcium chloride

Two chlorine atoms are required for each calcium atom as each calcium atom loses 2 electrons and each chlorine atom gains 1 electron. A calcium ion is Ca^{2+} as it has 18 electrons but it has 20 protons (atomic number = 20).

Chloride ions are Cl^- as described on page 24.

The formula of calcium chloride is $CaCl_2$.

Figure 3.6

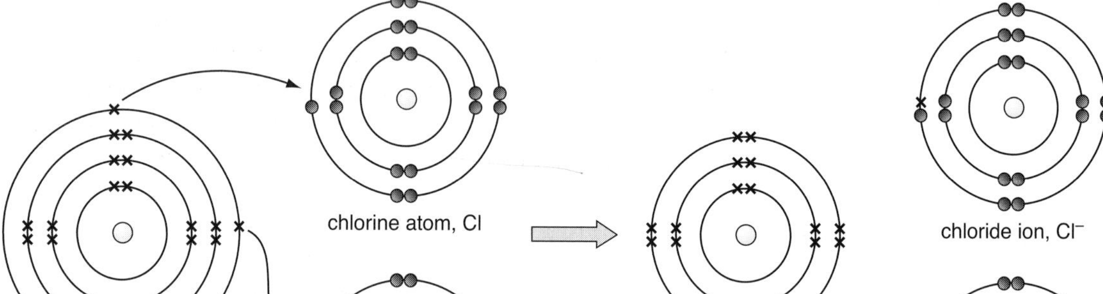

The ionic bond and properties of ionic compounds

The ionic bond

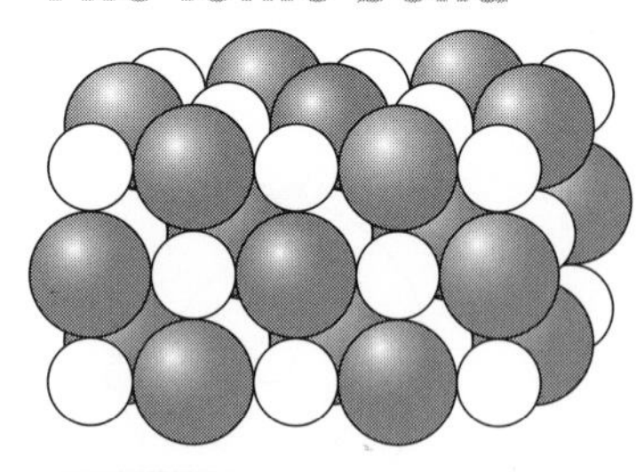

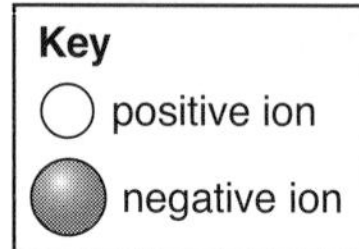

Figure 3.7 An ionic lattice

- An ionic bond is the attraction between oppositely charged ions.
- Ionic compounds have a lattice structure called an **ionic lattice**. Figure 3.7 shows one representation of this.
- The ionic lattice is a three-dimensional network of positive and negative ions held together by ionic bonds between the ions.
- All ionic bonds are strong and it is these bonds that give ionic compounds their properties.

Properties of ionic compounds

- In their solid state, ionic compounds cannot conduct electricity as the charged particles (which are the ions) cannot move and carry the charge.
- When an ionic compound is melted (it is described as **molten**) the ions can move and carry charge, so molten ionic compounds do conduct electricity (see page 53).
- Most ionic compounds dissolve in water.
- When an ionic compound dissolves in water (it is described as **aqueous**) the ions can move and carry charge, so aqueous ionic compounds do conduct electricity.
- Ionic compounds have high melting points and are all solids at room temperature and pressure, as a lot of energy is required to break the strong ionic bonds between the ions.

HINT: Learn the properties of ionic compounds carefully. A question can ask you to identify the type of an unknown compound just from its properties. If it has a high melting point and does not conduct electricity when solid but does when molten or when it is dissolved in water, it is an ionic compound.

Covalent bonding

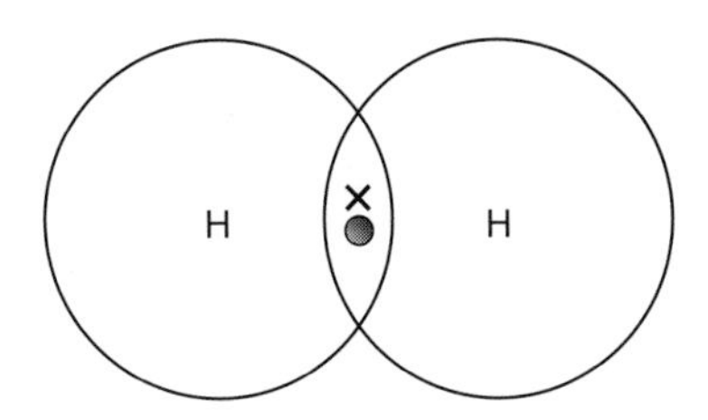

Figure 3.8 A covalent bond in hydrogen, H_2

A **covalent bond** occurs between non-metal atoms. It is formed from a **shared pair of electrons**. H_2 molecules contain a covalent bond between the hydrogen atoms, caused by a shared pair of electrons. One electron is from the H atom on the left and one from the H atom on the right as shown in Figure 3.8.

A covalent bond in hydrogen can be shown by 'H—H', where the long dash represents the bond. If two pairs of electrons are shared between non-metal atoms, this is a **double covalent bond**. A double covalent bond is represented by a double dash, 'O=O'.

Non-metal atoms share electrons to complete their outer shell.

NOTE: It is important to understand that the shared electrons count as outer shell electrons for *both* atoms.

Some more examples of covalent bonding are shown in Figure 3.9.

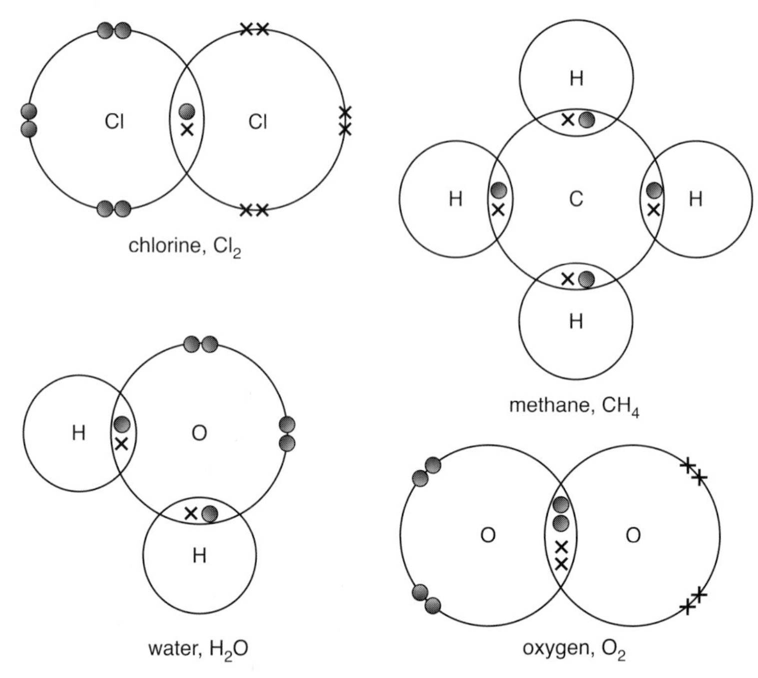

Figure 3.9 Electron diagrams showing covalent bonding

Diagrams of covalent bonding only need to show outer shell electrons, as in Figure 3.9. Molecules like the ones in Figure 3.9 are called simple covalent molecules and their structure is described as **molecular** or **simple**. Chlorine and oxygen are examples of simple covalent elements. They are **diatomic**. Water and methane are examples of simple covalent compounds.

Properties of simple covalent substances

Simple covalent substances have weak forces of attraction between molecules. The physical properties of simple covalent substances depend on these weak forces of attraction between the molecules. Their physical properties are as follows.

- They have low melting and boiling points because little energy is required to break the weak forces of attraction between the molecules.
- They do not conduct electricity as they have no charged particles (electrons or ions) which can move and carry charge.
- They are gases, liquids or low-melting point solids.

Giant covalent structures

- Some covalently bonded substances have **giant structures**.
- The giant covalent substances are diamond, graphite and quartz.
- Diamond and graphite are both forms of the element carbon.
- Alternative forms of the same element in the same physical state are called **allotropes**.
- Diamond and graphite are allotropes of carbon (Figure 3.10).

The structure of quartz is shown in Figure 3.11.

Figure 3.10 The giant structures of the allotropes, diamond and graphite

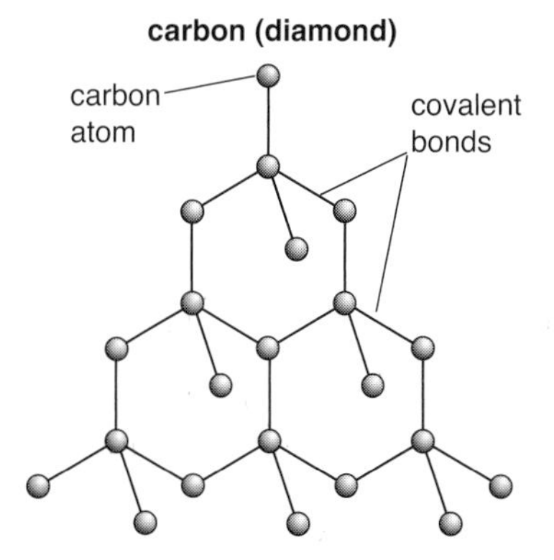

Each carbon atom in diamond is joined to four other carbon atoms by strong covalent bonds. The basic arrangement of bonds is called **tetrahedral**.

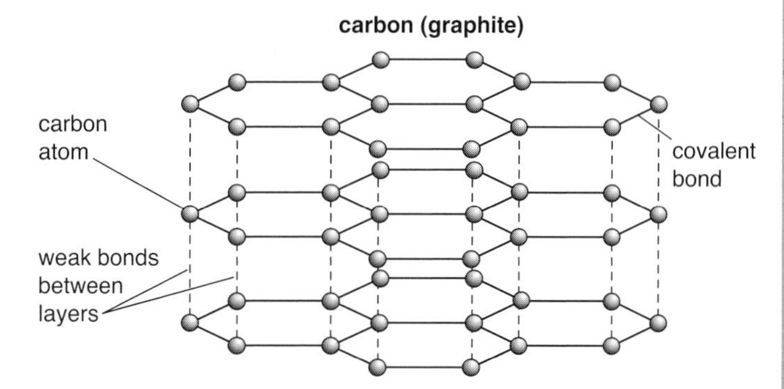

Each carbon atom in graphite is joined to three other carbon atoms by strong covalent bonds. The atoms are bonded in layers with weak bonds between the layers. The weak bonds between the layers are caused by **delocalised** electrons.

Figure 3.11 The giant structure of quartz

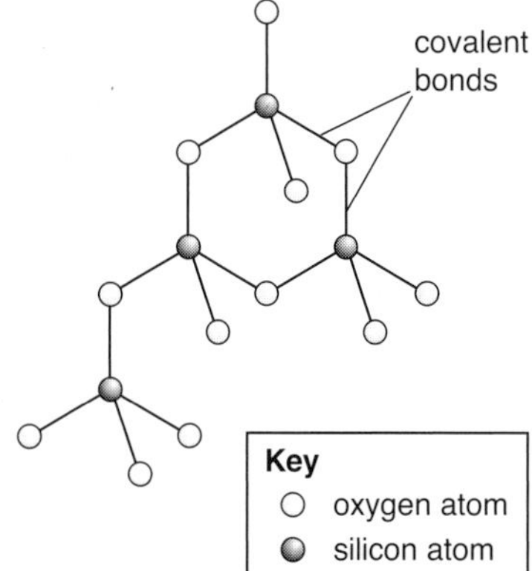

Each silicon atom in quartz is covalently bonded to four oxygen atoms and each oxygen atom is bonded to two silicon atoms in a tetrahedral arrangement.

HINT: Make sure you have practised drawing the structures of diamond, graphite and quartz. Being asked to draw the structures is a common question and valuable marks can be obtained by a correct diagram.

Typical question

Explain, in terms of bonds, why the reaction between hydrogen and chlorine producing hydrogen chloride is exothermic. [5]

Answer

The energy required to break the bonds [1] in hydrogen and chlorine [1] is less than [1] the energy released when bonds are made [1] in hydrogen chloride [1].

HINT: The most common mistake in this question is to forget to include the names of the reactants and products. If the reaction had been endothermic, the word 'less' is simply changed to 'more'.

Revision questions

1 What is meant by the term 'atomic number'? *[1]*

2 State the relative mass and relative charge of:
a a proton **b** an electron **c** a neutron. *[3]*

3 ^{35}Cl and ^{37}Cl are isotopes of chlorine.
What is meant by the term 'isotope'? *[1]*

4 Draw and write electronic configurations for the following atoms:
a P **b** Li **c** O **d** K
e Ar **f** He **g** Al **h** Na *[8]*

5 Write electronic configurations for the following ions:
a Na^+ **b** F^- **c** Al^{3+} **d** O^{2-} **e** K^+ *[5]*

6 Sodium chloride conducts electricity when it is molten but not when it is a solid. Explain why. *[3]*

7 Explain how a covalent bond forms. *[3]*

8 Draw a diagram to show the bonding in the following molecules. Only show outer shell electrons in your diagram.
a Cl_2 *[2]*
b H_2O *[4]*

9 Chlorine boils at −34 °C. Explain why chlorine has a low boiling point. *[3]*

10 Draw a diagram to show the structure of diamond. Describe the main features of the structure. *[3]*

11 Explain why graphite conducts electricity. *[3]*

12 What is meant by the term 'malleable'? *[1]*

13 Describe the structure and bonding in a metal such as sodium. *[3]*

14 Draw a labelled diagram of an atom of calcium showing the numbers of its subatomic particles. *[4]*

15 Copy and complete the table below for each of the particles X, Y and Z. The particles may be atoms or ions. The letters X, Y and Z do not represent symbols for elements.

Particle	Atomic number	Mass number	Number of protons	Number of neutrons	Number of electrons	Electronic configuration
X			11	12		2, 8
Y	17			18	17	
Z		16	8		10	

[3]

4

Formulae and equations

Valency

Table 4.1

Group	Valency
I	1
II	2
III	3
IV	4
V	3
VI	2
VII	1
0	–

To be able to work out the **formula** of a compound, you need to use **valency**. Valency is the combining power of an element or ion. It can be worked out from the group number of the elements as shown in Table 4.1.

Note that elements in Group 0 do not form compounds, so these have no valency. Hydrogen always has a valency of 1. The valency is also the same as the charge on the ions. The way in which the atoms become ions is discussed on page 24.

Molecular ions have the valencies shown in Table 4.2. These are the same as the charges on the molecular ions.

Table 4.2 Using charge on the molecular ion to work out valency

Name	Formula	Valency
ammonium	NH_4^+	1
sulphate	SO_4^{2-}	2
sulphite	SO_3^{2-}	2
carbonate	CO_3^{2-}	2
nitrate	NO_3^-	1
nitrite	NO_2^-	1
hydrogen carbonate	HCO_3^-	1
hydroxide	OH^-	1

HINT: For unfamiliar ions, the valency is the same as the charge. For example the thiosulphate ion, $S_2O_3^{2-}$, has a valency of 2, and the Cu^+ ion has a valency of 1.

The valency of the **transition metals** (middle block in the Periodic Table – see page 10) is given in roman numerals after the name of the element (see Table 4.3 overleaf).

Table 4.3 Using roman numerals in transition metal compounds to find valency

Transition metal compound	Valency of metal
iron(III) compounds	3 (iron)
iron(II) compounds	2 (iron)
copper(II) compounds	2 (copper)
cobalt(II) compounds	2 (cobalt)
copper(I) compounds	1 (copper)

HINT: Transition metal ions without a roman numeral usually have a valency of 2. The copper in copper sulphate has a valency of 2; the zinc in zinc chloride has a valency of 2. This is important as some examination questions do not always give the roman numeral.

How to work out the formula of a compound

➔ The formula of a compound can be determined using valencies by the following method.

1 Compound name — potassium chloride

2 Convert to symbols (or formula if molecular ion) — K Cl

3 Write valencies (using the group table or molecular ion table or transition metal numeral methods – see page 33 and above) — 1 (Group I) 1 (Group VII)

4 Write valencies above symbols — $\overset{1}{K}$ $\overset{1}{Cl}$

5 Cancel down if necessary (not needed for KCl, but for example: $\overset{2}{Mg}$ $\overset{2}{O}$ becomes $\overset{1}{Mg}$ $\overset{1}{O}$)

NOTE: This is the step most often forgotten.

6 Cross-over — $\overset{1}{K}$ ⇄ $\overset{1}{Cl}$

7 Write together with crossed-over numbers (use brackets if a molecular ion is multiplied by 2 or more) — K_1Cl_1

8 Ignore '1's (as $K_1 = K$) — KCl

Example 1

NOTE: Hydrogen, H, always has a valency of 1.

Use the valency method to work out the formulae of the following compounds.

1 Magnesium chloride

	Mg	Cl	
valency	2	1	
	^{2}Mg ←	→ ^{1}Cl	No cancel Cross-over
	Mg_1Cl_2		**$MgCl_2$**

2 Calcium oxide

	Ca	O	
valency	2	2	
	^{2}Ca	^{2}O	Cancel down
	^{1}Ca ←	→ ^{1}O	Cross-over
	Ca_1O_1		**CaO**

3 Iron(III) chloride

	Fe	Cl	
valency	3	1	
	^{3}Fe ←	→ ^{1}Cl	No cancel Cross-over
	Fe_1Cl_3		**$FeCl_3$**

4 Calcium hydroxide

	Ca	OH	
valency	2	1	
	^{2}Ca ←	→ ^{1}OH	No cancel Cross-over
	$Ca_1(OH)_2$		**$Ca(OH)_2$**

Covalent compounds are compounds containing only non-metals. We cannot use ions to determine the formulae of covalent compounds but the rules of valency still apply, as shown in Example 2.

Example 2

Use the valency method to work out the formulae of the following covalent compounds.

	Valencies	Formula
1 Hydrogen chloride	H (1), Cl (1)	**HCl**
2 Methane (C and H)	C (4), H (1)	**CH_4**
3 Ammonia (N and H)	N (3), H (1)	**NH_3**
4 Water (H and O)	H (1), O (2)	**H_2O**
5 Sulphuric acid (H and SO_4)	H (1), SO_4 (2)	**H_2SO_4**
6 Nitric acid (H and NO_3)	H (1), NO_3 (1)	**HNO_3**

HINT: The formulae of some compounds do not need working out. For example carbon dioxide is CO_2 as the 'di' indicates two oxygen atoms in the compound. Sulphur trioxide is SO_3. As you work through the examples, you will get quicker at working out formulae and it will become like a new language to you – the language of Chemistry!

Word equations and balanced symbol equations

➔ **Word equations** are a long way of representing reactions. **Balanced symbol equations** are a more convenient method of doing this.

Example 1

Write the balanced symbol equation for the reaction of sodium hydroxide with hydrochloric acid.

1 Write down the word equation for the reaction.

sodium hydroxide (aq) + hydrochloric acid (aq) → sodium chloride (aq) + water (l)

NOTE: Sometimes **state symbols** are given to show the physical state of each reactant and product. The symbols below are used to indicate this:

(aq) means aqueous (dissolved in water) **(l)** means liquid

(s) means solid **(g)** means gas

2 Now write the formulae of the reactants and the products.

sodium hydroxide	NaOH (aq)	reactants – substances which react together
hydrochloric acid	HCl (aq)	
sodium chloride	NaCl (aq)	products – substances produced in the reaction
water	H_2O (l)	

3 Substitute the formulae in the word equation to give you the symbol equation.

$$NaOH\ (aq) + HCl\ (aq) \rightarrow NaCl\ (aq) + H_2O\ (l)$$

4 It is important to **balance** a symbol equation so that there are the same numbers of each type of atom on both sides of the equation arrow.

In the above equation:

Atoms involved	Reactants side (left-hand side, LHS)	Products side (right-hand side, RHS)
Na	1	1
O	1	1
H	2	2
Cl	1	1

It can be seen that there is the same number of each type of atom on both sides of the equation. This symbol equation is already balanced.

So, the balanced symbol equation is:

$$NaOH\ (aq) + HCl\ (aq) \rightarrow NaCl\ (aq) + H_2O\ (l)$$

NOTE: The state symbols are not always necessary and you should only include them when asked to do so in a question. The equation can be written:

$$NaOH + HCl \rightarrow NaCl + H_2O$$

HINT: Be careful to use an arrow and not an equals sign in your equations – this is a chemical equation, not a mathematical one!

Example 2

Write the balanced symbol equation for the reaction of magnesium with hydrochloric acid.

1 Write down the word equation for the reaction.

magnesium + hydrochloric acid → magnesium chloride + hydrogen

2 Formulae:

magnesium	Mg	reactants – substances which react together
hydrochloric acid	HCl	
magnesium chloride	$MgCl_2$	products – substances produced in the reaction
hydrogen	H_2	

3 Symbol equation: $Mg + HCl \rightarrow MgCl_2 + H_2$

4 Balance the equation:

Atoms involved	Reactants side (left-hand side, LHS)	Products side (right-hand side, RHS)
Mg	1	1
H	1	2
Cl	1	2

It can be seen clearly that the above symbol equation is not balanced but *it is important to remember that formulae cannot be changed to balance an equation.*

To balance an equation, balancing numbers are written *in front* of specific formulae and the whole formula then becomes multiplied by this number. For example, '2HCl' balances the above equation as it gives us 2 H atoms and 2 Cl atoms on the left-hand side.

So, the balanced symbol equation is:

$$Mg + 2HCl \rightarrow MgCl_2 + H_2$$

Example 3

Write the balanced symbol equation for the reaction of magnesium hydroxide with nitric acid.

1 Write down the word equation for the reaction.

magnesium hydroxide + nitric acid → magnesium nitrate + water

2 Formulae in symbol equation:

$$Mg(OH)_2 + HNO_3 \rightarrow Mg(NO_3)_2 + H_2O$$

3 For balancing purposes, the nitrate, NO_3, stays intact (i.e. does not break up) in this reaction and so can be considered as a unit. (However, if the molecular ion does break up, the atoms must be considered separately.)

Atoms involved	Reactants side (left-hand side, LHS)	Products side (right-hand side, RHS)
Mg	1	1
O (ignoring the 'O' in nitrate)	2	1
H	3	2
NO_3	1	2

This requires balancing:

- We need 2 NO_3 on LHS, so $2HNO_3$ are needed.
- Now 4 H on LHS and 2 H on RHS so $2H_2O$ are needed.
- Oxygen is now balanced.

So, the balanced symbol equation is:

$$Mg(OH)_2 + 2HNO_3 \rightarrow Mg(NO_3)_2 + 2H_2O$$

Example 4

Write the balanced symbol equation for the reaction of calcium carbonate with hydrochloric acid.

1 Write down the word equation for the reaction.

calcium carbonate + hydrochloric acid → calcium chloride + carbon dioxide + water

2 Formulae in symbol equation:

$$CaCO_3 + HCl \rightarrow CaCl_2 + CO_2 + H_2O$$

3 Balance the equation:

Atoms involved	Reactants side (left-hand side, LHS)	Products side (right-hand side, RHS)
Ca	1	1
C	1	1
O	3	3
H	1	2
Cl	1	2

It can be seen clearly that there are 2 H atoms and 2 Cl atoms on the RHS, while the LHS has only 1 of each. Therefore, 2HCl are needed to balance.

Table 5.2 gives the expected colours with Universal indicator for strong acids, weak acids, neutral solution, weak alkalis and strong alkalis. Common examples are also given.

Table 5.2 The colours of Universal indicator in different conditions

<table>
<tr><th>pH</th><th>0</th><th>1</th><th>2</th><th>3</th><th>4</th><th>5</th><th>6</th><th>7</th><th>8</th><th>9</th><th>10</th><th>11</th><th>12</th><th>13</th><th>14</th></tr>
<tr><td>colour with Universal indicator</td><td colspan="3">red</td><td colspan="2">orange</td><td colspan="2">yellow</td><td>green</td><td colspan="2">green–blue</td><td colspan="2">blue</td><td colspan="3">purple</td></tr>
<tr><td>strength</td><td colspan="3">strong acid</td><td colspan="4">weak acid</td><td>neutral</td><td colspan="4">weak alkali</td><td colspan="3">strong alkali</td></tr>
<tr><td>examples</td><td colspan="3">HCl
H_2SO_4
HNO_3</td><td colspan="4">CH_3COOH
(citric acid)</td><td>water</td><td colspan="4">NH_3</td><td colspan="3">NaOH
KOH</td></tr>
<tr><td>common solutions</td><td colspan="3">gastric juice</td><td colspan="4">vinegar
orange juice</td><td>pure water</td><td colspan="4">blood
seawater</td><td colspan="3">washing soda solution</td></tr>
</table>

Ions in acids and alkalis

- All acids dissolve in water producing hydrogen ions in solution, H^+ (aq).
- All alkalis dissolve in water producing hydroxide ions in solution, OH^- (aq).
- **Neutralisation** is the reaction between an acid and an alkali producing a salt and water only. It can be represented by:

  acid + alkali → salt + water

- The equation for neutralisation can be written as an **ionic equation** (see page 41):

  $$H^+ (aq) + OH^- (aq) \rightarrow H_2O(l)$$

Typical question

Write an ionic equation for neutralisation including state symbols. *[2]*

Answer

$H^+ (aq) + OH^- (aq) \rightarrow H_2O (l)$ *[2]*

HINT: The main problem with this type of question is that the state symbols are left out altogether, or that all of the state symbols are written as '(aq)'.

Bases and alkalis

- A **base** is a substance that reacts with an acid producing a salt and water only.
- Common bases are metal oxides and metal hydroxides. Copper(II) oxide, CuO; magnesium oxide, MgO; potassium hydroxide, KOH and sodium hydroxide, $NaOH$ are all bases as they are metal oxides or hydroxides.
- An **alkali** is a soluble base.
- The most common alkalis are sodium hydroxide, $NaOH$; potassium hydroxide, KOH; calcium hydroxide, $Ca(OH)_2$ and ammonia, NH_3.
- Ammonia is not a metal oxide or an hydroxide but a solution of ammonia contains hydroxide ions, OH^- so it is an alkali.
- Ammonia reacts with water to produce a solution of ammonium hydroxide:

$$NH_3 + H_2O \rightarrow NH_4OH$$

HINT: Check the back of the *Data Leaflet* for solubility information if you are not sure. The majority of oxides and hydroxides are insoluble in water, so most are bases and only a few can also be called alkalis. Don't forget that ammonia is an alkali.

Nature of oxides

Look back at page 10 to remind yourself of the positions of the elements in the Periodic Table.

Oxides of elements can be basic, acidic, neutral (in a few cases) or amphoteric. The reactions of oxides can be summarised as follows.

1. **Basic oxides** (bases) are generally oxides of metals that react with acids to produce a salt and water only, for example MgO and CuO.
2. **Acidic oxides** are generally oxides of non-metals that will react with alkalis to form a salt and water only, for example CO_2 and SO_2. (See also page 130.)

$$CO_2 + 2NaOH \rightarrow Na_2CO_3 + H_2O$$

3. **Neutral oxides** form neutral solutions, for example H_2O and CO.

4 **Amphoteric oxides** are oxides of metals that react with both acids and alkalis. The reactions of aluminium oxide, Al_2O_3, and zinc oxide, ZnO, are the ones you need to know.

- Reaction of Al_2O_3 with an acid:
  $Al_2O_3 + 6HCl \rightarrow 2AlCl_3 + 3H_2O$
- Reaction of Al_2O_3 with an alkali:
  $Al_2O_3 + 2NaOH \rightarrow 2NaAlO_2 + H_2O$
- Reaction of ZnO with an acid:
  $ZnO + 2HCl \rightarrow ZnCl_2 + H_2O$
- Reaction of ZnO with an alkali:
  $ZnO + 2NaOH \rightarrow Na_2ZnO_2 + H_2O$
- $NaAlO_2$ is called sodium aluminate;
  Na_2ZnO_2 is called sodium zincate.
- The aluminate ion is AlO_2^- and the zincate ion is ZnO_2^{2-}.

HINT: You need to learn the equations for the reactions of the amphoteric oxides with alkali. They can be simplified to an ionic equation that is suitable for the reaction with any alkali:

$Al^{3+} + 4OH^- \rightarrow [Al(OH)_4]^-$

and $Zn^{2+} + 4OH^- \rightarrow [Zn(OH)_4]^{2-}$

These two ionic equations will be accepted for the reaction of the amphoteric oxide (represented by Al^{3+} or Zn^{2+}) with an alkali (OH^-).

Reactions of acids

- Acids produce salts when they react.
- Acids are solutions containing hydrogen ions, H^+, and a negative ion.
- The negative ion (called an anion) is what combines with a metal ion to form the salt.
- Look at page 105 for more on reactions of metals with acids.

Hydrochloric acid, HCl, contains hydrogen ions, H^+, and **chloride** ions, Cl^-.

So when hydrochloric acid reacts, it forms a **chloride**.

Sulphuric acid, H_2SO_4, contains hydrogen ions, H^+, and **sulphate** ions, SO_4^{2-}.

So when sulphuric acid reacts, it forms a **sulphate**.

Ethanoic acid, CH_3COOH, contains hydrogen ions, H^+, and **ethanoate** ions, CH_3COO^-.

When ethanoic acid reacts, it forms an **ethanoate**.

Remove the H^+ ions from an acid formula to work out the formula of the anion. For every H^+ you remove, give the anion one more negative charge.

The reactions of acids can be summarised as follows.

1 **Acids + metals:** dilute acids react with some metals to produce a salt, and hydrogen gas is released. (Remember that dilute acids do not react with copper metal or any other metal below copper in the reactivity series – see page 109 for the reactivity series.)

general reaction: metal + acid → salt + hydrogen

Example:

magnesium + hydrochloric acid → magnesium chloride + hydrogen

$$Mg + 2HCl \rightarrow MgCl_2 + H_2$$

NOTE: Don't forget that hydrogen is diatomic.

2 **Acid + metal oxides/hydroxides:** dilute acids react with metal oxides and metal hydroxides to produce a salt and water. Metal oxides and hydroxides are bases.

general reactions: metal oxide + acid → salt + water
metal hydroxide + acid → salt + water

Examples:

copper(II) oxide + sulphuric acid → copper sulphate + water

$$CuO + H_2SO_4 \rightarrow CuSO_4 + H_2O$$

sodium hydroxide + hydrochloric acid → sodium chloride + water

$$NaOH + HCl \rightarrow NaCl + H_2O$$

3 **Acid + metal carbonate/hydrogen carbonate:** dilute acids react with metal carbonates and metal hydrogen carbonates to produce a salt, carbon dioxide and water.

general reactions:
metal carbonate + acid → salt + carbon dioxide + water
metal hydrogen carbonate + acid → salt + carbon dioxide + water

Examples:

sodium carbonate + sulphuric acid → sodium sulphate + carbon dioxide + water

$$Na_2CO_3 + H_2SO_4 \rightarrow Na_2SO_4 + CO_2 + H_2O$$

potassium hydrogen carbonate + nitric acid → potassium nitrate + carbon dioxide + water

$$KHCO_3 + HNO_3 \rightarrow KNO_3 + CO_2 + H_2O$$

4 **Acid + ammonia:** dilute acids react with ammonia to form an ammonium salt only.

general reaction: ammonia + acid → ammonium salt

Examples:

ammonia + hydrochloric acid → ammonium chloride
$NH_3 + HCl \rightarrow NH_4Cl$

ammonia + sulphuric acid → ammonium sulphate
$2NH_3 + H_2SO_4 \rightarrow (NH_4)_2SO_4$

Preparing a salt

When preparing a pure dry sample of a salt there are three different methods, depending on the solubility of the salt that is being prepared and the solubility of the starting compounds.

Making a pure, dry sample of a soluble salt is divided into preparation and purification.

In all salt preparations, the acid is chosen based on the anion in the salt:

- if the salt is a chloride, hydrochloric acid is used
- if the salt is a sulphate, sulphuric acid is used
- if the salt is a nitrate, nitric acid is used
- if the salt is an ethanoate, ethanoic acid is used.

HINT: Remember the solubility of compounds is given on the back of the *Data Leaflet*. Use this to determine which of the following three methods to use.

Method 1: for all sodium, potassium or ammonium salts

This method is carried out by titration of an acid against an alkali (see page 84).

The alkali is either sodium hydroxide solution, potassium hydroxide solution or ammonia solution, depending on what positive ion is in the salt.

Titration: Pipette 25.0 cm^3 of alkali into a conical flask and add phenolphthalein indicator which gives a pink colour. Charge the burette with acid and titrate until the colour changes to colourless. Record the volume of acid used. Repeat without an indicator.

If an ammonium salt is being prepared, methyl orange indicator should be used; it will be yellow in the ammonia solution in the conical flask and will change to red.

Instead of repeating without indicator, the indicator may be removed by adding charcoal, heating and filtering the solution. The charcoal absorbs the indicator.

Purification: Heat to evaporate to reduce the volume by half and leave aside to cool and crystallise. Then filter off the crystals and dry between two sheets of filter paper *or* in a low-temperature oven *or* in a desiccator.

Method 2: for all soluble salts (except those in method 1)

This method requires an insoluble solid containing the metal cation.

- Metal oxides, hydroxides and carbonates are mostly insoluble.
- The metal carbonate has advantages because when it reacts it produces a gas (CO_2) so it is easier to tell when the solid is in excess as no more gas will be produced.
- If an oxide or hydroxide is used, the solid is in excess when some of it lies at the bottom unreacted.
- By adding excess solid, the acid is all used up, which leaves a pure solution of the salt after filtration.

Preparation: Add excess of the insoluble solid to 25 cm^3 of acid in a conical flask. Heat and stir and then filter off the insoluble solid (**residue**) and pour the filtered liquid (**filtrate**) into an evaporating basin.

Purification: Heat to evaporate to reduce the volume by half and leave aside to cool and crystallise. Then filter off the crystals and dry between two sheets of filter paper *or* in a low-temperature oven *or* in a desiccator.

Typical question

Zinc carbonate is added to dilute sulphuric acid until it is in excess. Describe how you would know the zinc carbonate was in excess. *[2]*

Answer

No more [1] gas [1] produced or solid [1] remains at the bottom [1].

HINT: Questions on the practical aspects of chemistry are never well answered. This type of question is typical. Always look at a reaction of an acid to see if a gas is produced (CO_2 if the acid is reacting with a carbonate or a hydrogen carbonate; H_2 if the acid is reacting with a metal). If it is not producing a gas, the solid remaining at the bottom will ensure all the acid is used up.

Method 3: for all insoluble salts

- To make an insoluble salt such as barium sulphate, two soluble compounds are chosen one of which contains the **cation** (for example, barium ion) and one of which contains the **anion** (for example, sulphate ion).
- From the back of the *Data Leaflet*, you can see that barium chloride and sodium sulphate are soluble.
- Use solutions of these two and when they are combined a precipitate is formed.
- A **precipitate** is a solid produced in a solution during a reaction. 'Precipitate' can be abbreviated to 'ppt'.

Preparation: Mix the two solutions. A precipitate forms.

Purification: Filter off the precipitate and wash it with water. Then dry between two sheets of filter paper *or* in a low-temperature oven *or* in a desiccator.

The colour of the precipitate should be stated. For example, for barium sulphate you should say 'a white precipitate is formed'.

HINT: Often the ionic equation for the precipitation of an insoluble salt is asked for in a question. Using barium sulphate as our example, the balanced symbol equation is:

$BaCl_2\ (aq) + Na_2SO_4\ (aq) \rightarrow BaSO_4\ (s) + 2NaCl\ (aq)$

The ionic equation is:

$Ba^{2+}\ (aq) + SO_4^{2-}\ (aq) \rightarrow BaSO_4\ (s)$

(The Na^+ and Cl^- ions do not take part in the reaction.)

(See also page 41.)

Anion tests

The anion present in salts can be tested for using a variety of methods. For the chloride, bromide, iodide and sulphate ions, a solution of the compound should be made first and then the second solution should be added (Table 5.3).

Table 5.3 How to test for an anion present in a salt

Testing for anion	Add second solution	Result
carbonate	dilute hydrochloric acid	colourless, odourless gas
chloride	silver nitrate solution	white ppt
bromide	silver nitrate solution	cream ppt
iodide	silver nitrate solution	yellow ppt
sulphate	barium chloride solution	white ppt

HINT: Again a question can ask for ionic equations for all the precipitations. For example, the addition of silver nitrate to a solution of sodium bromide will produce a cream precipitate.

The balanced symbol equation for this reaction is:

$AgNO_3$ (aq) + NaBr (aq) → AgBr (s) + $NaNO_3$ (aq)

The ionic equation for the reaction is:

Ag^+ (aq) + Br^- (aq) → AgBr (s)

Silver bromide, AgBr, is the cream solid that forms as the precipitate in this reaction.

Revision questions

1 What is an alkali? [2]

2 Name **two compounds** that would react with sulphuric acid to prepare a solution of copper sulphate. [2]

3 Write an equation for the reaction of aluminium oxide with potassium hydroxide. [3]

4 What would be observed when a solution of silver nitrate is added to a solution containing potassium iodide? [2]

5 Write the formula of the ion which is present in all acids. [1]

6 Name **two** acidic oxides. [2]

7 Name **two** solutions which could be used to prepare a sample of barium sulphate. [2]

8 Write an ionic equation for the reaction of silver nitrate solution with sodium chloride solution. [2]

9 State the colours observed when phenolphthalein is added to:
- **a** hydrochloric acid
- **b** water
- **c** sodium hydroxide solution [3]

10 State the colours observed when methyl orange is added to each of the solutions listed in question **9**. [3]

11 Write a balanced symbol equation for each of the following reactions of hydrochloric acid:
- **a** magnesium + hydrochloric acid [3]
- **b** magnesium hydroxide + hydrochloric acid [3]
- **c** sodium carbonate + hydrochloric acid [3]

12 What salt is produced when potassium hydroxide solution reacts with sulphuric acid? [1]

13 What gas is produced when zinc reacts with dilute sulphuric acid? [1]

14 What is the approximate pH of the following solutions?
- **a** vinegar
- **b** water
- **c** sodium hydroxide solution [3]

15 Copper carbonate is added in excess to dilute hydrochloric acid. The excess copper carbonate is filtered off and the filtered solution is heated to reduce the volume of water. On cooling, crystals are obtained.
- **a** What is the general name given to the solid which is removed by filtration? [1]
- **b** What is the general name given to the filtered solution? [1]
- **c** Explain how the crystals are removed from the solution and dried. [2]

6

Electrolysis

Defining electrolysis

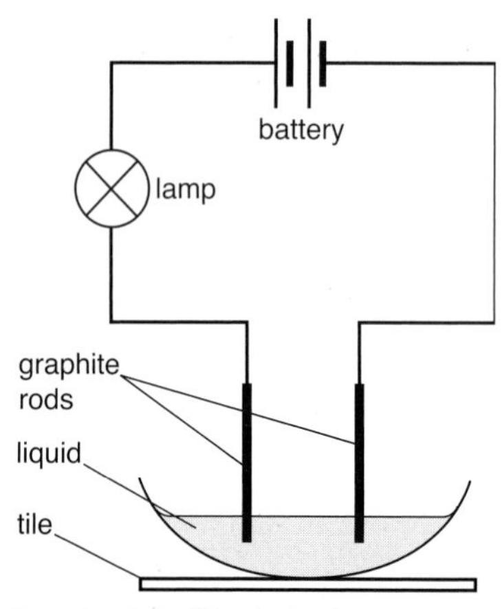

Figure 6.1 The apparatus used to test if a liquid conducts electricity

Two graphite rods, placed in a liquid and connected externally to a power supply such as a battery or a power pack, can be used to test if a liquid conducts electricity (Figure 6.1).

If the liquid conducts electricity and is **decomposed** by it, then **electrolysis** is taking place.

> Electrolysis is the decomposition of a liquid electrolyte using a direct current of electricity.

The **electrolyte** is the liquid or solution which conducts electricity and is decomposed by it. The graphite rods used are called **electrodes**. Graphite is used because it conducts electricity and is unreactive. However, in some electrolysis reactions other materials are used which are more **inert** (unreactive) such as titanium or platinum.

HINT: The definitions of the terms used in electrolysis are asked in almost every electrolysis question. Learn them exactly as they are written above. The reasons why graphite or titanium or platinum are used for the electrodes are: 'they conduct electricity and they are unreactive'. This is a frequently asked question.

How electrolysis works

All electrolytes conduct electricity as they have **free ions** that can move and carry charge. When these positive and negative ions are free to move, the positive ions (called **cations**) move to the **negative electrode** (called the **cathode**) and the **negative ions** (called **anions**) move to the positive electrode (called the **anode**).

- The positive ions at the negative electrode *gain* electrons to become atoms (which may combine to form diatomic molecules in the case of H_2).
- The negative ions at the positive electrode *lose* electrons to become atoms (which may combine to form diatomic molecules for all other diatomics).

Types of electrolytes

There are three types of electrolytes that you need to know with the following examples for each:

- molten ionic compounds – lithium chloride (LiCl), lead bromide ($PbBr_2$)
- aqueous ionic compounds – copper(II) sulphate ($CuSO_4$), sodium chloride (NaCl)
- acids – dilute sulphuric acid (H_2SO_4).

Observations and half equations

For each of the following electrolytes, you must be able to describe what is observed at each of the electrodes. You must also be able to write **half equations** to represent the reactions at each of the electrodes in terms of loss or gain of electrons.

Molten LiCl

anode: observations: yellow-green pungent gas evolved
half equation: $2Cl^- \rightarrow Cl_2 + 2e^-$

cathode: observations: silvery grey liquid formed
half equation: $Li^+ + e^- \rightarrow Li$

Molten $PbBr_2$

anode: observations: red-brown pungent gas evolved
half equation: $2Br^- \rightarrow Br_2 + 2e^-$

cathode: observations: silvery grey liquid formed which sinks to the bottom (can only be seen when molten electrolyte poured off)
half equation: $Pb^{2+} + 2e^- \rightarrow Pb$

Aqueous NaCl

anode: observations: yellow-green pungent gas evolved
half equation: $2Cl^- \rightarrow Cl_2 + 2e^-$

cathode: observations: colourless odourless gas evolved
half equation: $2H^+ + 2e^- \rightarrow H_2$

This electrolysis of sodium chloride is the basis of the **chloralkali** industry. Three useful materials can be made by electrolysing seawater. The seawater is electrolysed and the products at the electrodes are hydrogen and chlorine.
The solution remaining is sodium hydroxide. The anode is made of titanium as other materials may react with chlorine.

Aqueous $CuSO_4$ (graphite electrodes)

anode: observations: colourless odourless gas evolved
half equation:
$4OH^- \rightarrow O_2 + 2H_2O + 4e^-$

cathode: observations: red-pink solid deposited
half equation:
$Cu^{2+} + 2e^- \rightarrow Cu$

electrolyte: observations: blue colour fades to colourless

Aqueous $CuSO_4$ (copper electrodes)

anode: observations: anode dissolves
half equation: $Cu \rightarrow Cu^{2+} + 2e^-$

cathode: observations: red-pink solid deposited
half equation:
$Cu^{2+} + 2e^- \rightarrow Cu$

electrolyte: observations: blue colour remains the same

The anode loses mass and the cathode gains mass during this electrolysis.

Dilute H_2SO_4

anode: observations: colourless, odourless gas evolved
half equation:
$4OH^- \rightarrow O_2 + 2H_2O + 4e^-$

cathode: observations: colourless odourless gas evolved
half equation:
$2H^+ + 2e^- \rightarrow H_2$

The electrodes are made from platinum in the electrolysis of sulphuric acid as sulphuric acid may react with other materials.

Typical question

Explain, in terms of ions, why molten lead bromide conducts electricity. *[3]*

Answer

The ions *[1]* are free to move *[1]* and carry charge *[1]*.

HINT: The answer must include mention of ions and movement. The most common mistake is to state 'carry current' instead of 'carry charge'.

Apparatus used to electrolyse dilute sulphuric acid

The apparatus shown in Figure 6.2 is used to electrolyse sulphuric acid and collect the gases evolved.

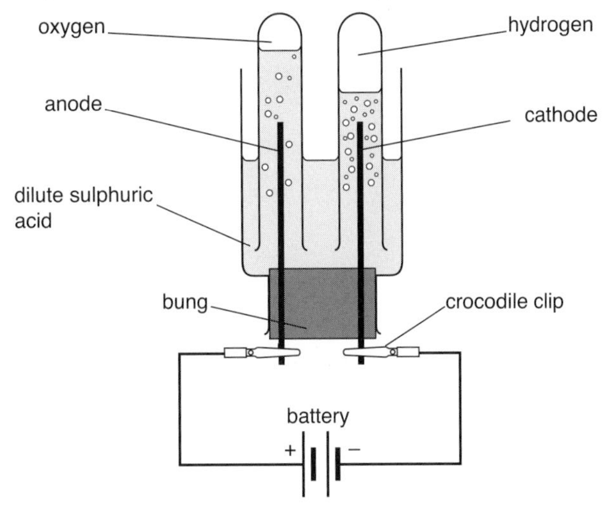

Figure 6.2

The volume of hydrogen is twice the volume of oxygen, as 2 electrons given up at the cathode produce 1 H_2 molecule; whereas it takes 4 electrons to be removed at the anode to produce 1 O_2 molecule.

Four electrons lost and gained will produce 2 H_2 molecules for every 1 O_2 molecule. The electrodes are usually made of platinum as it is very unreactive and conducts electricity.

Apparatus used to electrolyse a molten electrolyte

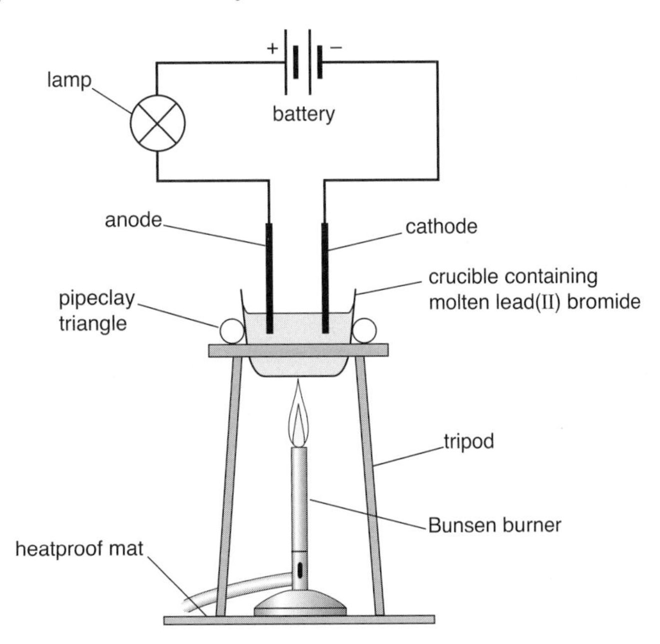

Figure 6.3

HINT: When you are asked in a question for a 'labelled diagram', marks are awarded for the labels only – if you have drawn the apparatus correctly assembled but not labelled you will get no marks.

Apparatus used to electrolyse a solution

The apparatus shown in Figure 6.4 is used to electrolyse a solution when gases are not being collected.

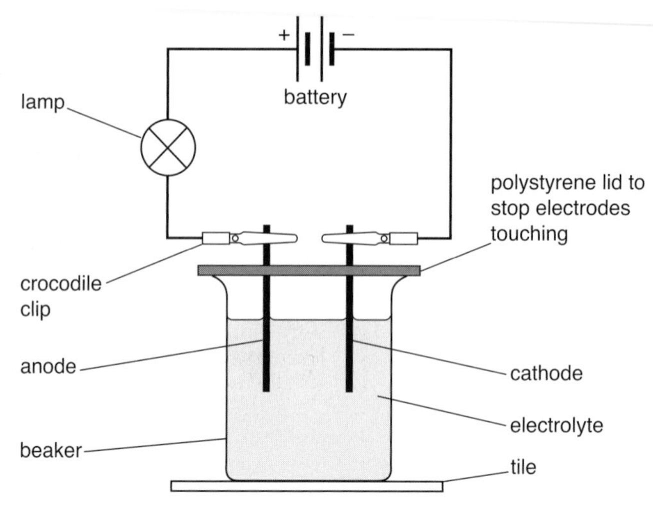

Figure 6.4

Extraction of aluminium

Aluminium metal is extracted from its ore using electrolysis. The ore is called **bauxite**. Bauxite is purified to form aluminium oxide (called **alumina**). The alumina is dissolved in molten **cryolite** to reduce its melting point and increase the conductivity.

The crust of aluminium oxide keeps heat in. The operating temperature is 900 °C.

The cathode and anode are made of carbon.

- The reaction at the cathode is: $Al^{3+} + 3e^- \rightarrow Al$
- The reaction at the anode is: $2O^{2-} \rightarrow O_2 + 4e^-$

The carbon anode has to be replaced periodically as it wears away because of its reaction with oxygen. The equation for this reaction is:

$$C + O_2 \rightarrow CO_2$$

Figure 6.5 shows the apparatus used to extract aluminium.

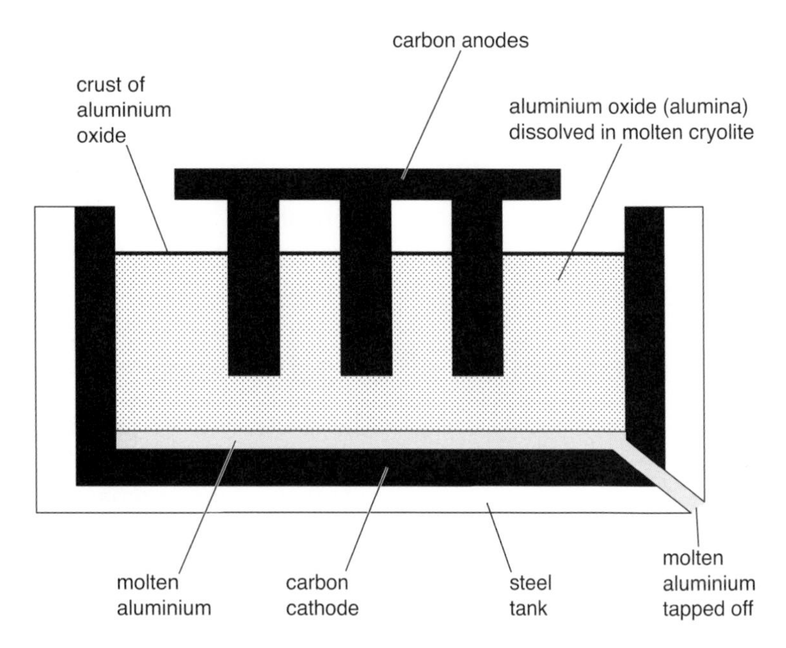

Figure 6.5 The apparatus used to extract aluminium from its ore

Refining of copper

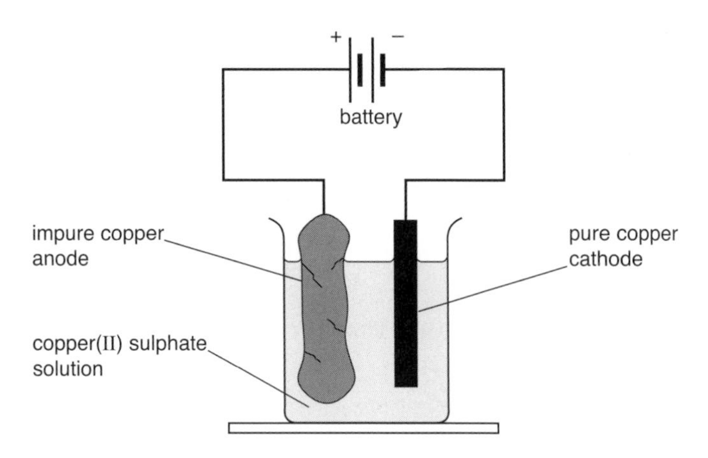

Figure 6.6 The apparatus used to refine copper

In Figure 6.6:

- **at the anode:** the impure copper anode dissolves:

  $Cu \rightarrow Cu^{2+} + 2e^-$

- **at the cathode:** the pure copper cathode has copper deposited on it:

  $Cu^{2+} + 2e^- \rightarrow Cu$

- **the electrolyte:** the concentration of the solution and its colour remain constant. The colour remains constant because for every Cu^{2+} ion that is removed at the cathode, it is replaced by a Cu^{2+} from the anode.

HINT: Always label diagrams carefully. The marks are for the labels.

The half equations are often called ionic equations. The half equations at the cathode involve the *addition* of electrons. The half equations at the anode involve the *loss* of electrons. Remember:

- when electrons are added, they go on the left of the half equation
- when electrons are lost, they go on the right of the half equation.

The economics of electrolysis

Any electrolytic extraction or refining (pages 56–57) requires large amounts of electricity. Electricity is a major cost factor for a company making aluminium. It makes the pure aluminium more expensive to buy as the company has to extract it electrochemically.

Revision questions

1 What is observed at the anode during the electrolysis of molten lead(II) bromide? *[3]*

2 Write a half equation for the reaction at the anode during the electrolysis of molten lithium chloride. *[3]*

3 What is meant by the term 'electrolysis'? *[2]*

4 What material is used for the electrodes during the extraction of aluminium? *[1]*

5 What do you understand by the following terms?
a the anode
b the cathode *[4]*

6 What does the term 'electrolyte' mean? *[3]*

7 What material is used for the anode and the cathode during the refining of copper? *[4]*

8 State **two** reasons why graphite is used for the electrodes during the electrolysis of molten lead(II) bromide. *[2]*

9 What is the name of aluminium ore? *[1]*

10 What material is alumina dissolved in before electrolysis? *[1]*

11 Write a half equation for the ionic reaction occurring at the cathode during the electrolysis of aqueous sodium chloride. *[3]*

12 Explain why the anode has to be replaced regularly during the electrolytic extraction of aluminium.
You may use an equation to help you answer the question. *[2]*

13 Write a balanced ionic equation for the discharge of bromide ions at the anode. *[3]*

14 What gas is produced at each electrode during the electrolysis of dilute sulphuric acid using platinum electrodes? *[2]*

15 Write an equation for the discharge of hydroxide ions at the anode during electrolysis. *[3]*

7

Water

Chemical tests for water

There are two main chemical tests for water.

1. **Anhydrous copper(II) sulphate** – white solid changes to blue.
2. **Anhydrous cobalt(II) chloride** (or cobalt chloride paper) – pale blue changes to pink.

Water pollution

- The main dissolved **pollutants** in water are **nitrates** and **phosphates**.
- Farmers use artificial fertilisers on their fields. Run-off from the fields contains nitrates that leach into rivers and lakes, causing **eutrophication**.
- Eutrophication is caused by an increased level of dissolved nitrates in water courses. This leads to increased growth of **algae**. Some algae die due to increased **competition**, and **decomposition** of dead algae uses up oxygen. Fish die due to lack of oxygen.
- **Nitrates** in drinking water can cause blue-baby syndrome, and are a possible cause of stomach cancer. Blue-baby syndrome occurs when insufficient oxygen gets to a baby during pregnancy. Nitrates in a person's stomach can be converted to cancer-causing chemicals.
- **Phosphates** from detergents and fertilisers can also enter the water courses. Phosphates may cause eutrophication, and high levels of phosphates in drinking water may cause digestive problems.
- Drinking water is filtered to remove insoluble substances and chlorinated to kill bacteria.

Hard water

- **Hard water** is water that does not readily produce a lather with soap. Water which lathers easily with soap is called **soft water**. Note that hard water will form a lather with detergents.
- Hardness in water is caused by dissolved **Ca^{2+} ions** or **Mg^{2+} ions**.
- There are two types of hardness in water – **temporary hardness** and **permanent hardness**.
- Temporary hardness can be removed by boiling.
- Permanent hardness cannot be removed by boiling.
- Temporary hardness is caused by dissolved calcium hydrogen carbonate.
- Permanent hardness is caused by dissolved calcium or magnesium sulphate, or dissolved calcium or magnesium chloride.

Testing for hardness

Table 7.1 shows the method of testing water for the two types of hardness – temporary and permanent.

Table 7.1 Method of testing water for the two types of hardness

Temporary hardness	Permanent hardness
1 take a sample of water and add soap	1 take a sample of water and add soap
2 shake it and there should be *no lather*	2 shake it and there should be *no lather*
3 take another sample of water and boil it	3 take another sample of water and boil it
4 add soap and shake	4 add soap and shake
5 there should be a *lather*	5 there should *still be no lather*

Hard water and soap

Ca^{2+} (or Mg^{2+}) ions in the hard water react with the stearate ions in soap producing an **insoluble scum/solid** of **calcium stearate** (or **magnesium stearate**).

How temporary hardness arises in water

Limestone ($CaCO_3$) reacts with rainwater containing dissolved carbon dioxide to form calcium hydrogen carbonate solution.

The equation for this reaction is:

$$CaCO_3 + H_2O + CO_2 \rightarrow Ca(HCO_3)_2$$

Removing hardness in water

Any method of removing hardness from water must remove the dissolved Ca^{2+} (or Mg^{2+}) ions. This can be achieved by removing them from solution by precipitating them out as an insoluble solid (by adding washing soda or by boiling) or by exchanging them for Na^{+} ions which do not cause hardness.

Boiling
Boiling removes only temporary hardness in water. Dissolved Ca^{2+} ions are removed as insoluble $CaCO_3$.

The equation for the reaction is:

$$Ca(HCO_3)_2 \rightarrow CaCO_3\,(s) + CO_2 + H_2O$$

HINT: This equation is the reverse of the formation of temporary hardness in water. This means you only have to learn one equation (for the formation of temporary hardness) and then reverse it for the removal of temporary hardness by boiling.

Ion exchange
Ion exchange removes both permanent and temporary hardness in water. Dissolved Ca^{2+} (or Mg^{2+}) ions in the hard water are removed and replaced by Na^{+} ions from the ion exchanger.

Washing soda (hydrated sodium carbonate)
Washing soda removes both permanent and temporary hardness in water. Dissolved Ca^{2+} (or Mg^{2+}) ions from the hard water are removed due to a reaction with carbonate ions from the washing soda to form insoluble $CaCO_3$ (or $MgCO_3$).

The equation for this reaction is:

$$Ca^{2+}\,(aq) + CO_3^{2-}\,(aq) \rightarrow CaCO_3\,(s)$$

Where hard water occurs

The characteristic features of hard water geographical regions are **stalagmites**, **stalactites** and **caves**.

Advantages and disadvantages of hard water

Advantages

- Hard water tastes better.
- It is better for brewing beer.
- It is good for tanning leather.
- It provides calcium (Ca^{2+}) for healthy bones and teeth.

Making a solid dissolve more quickly

To make a solid dissolve more quickly, several things can be done:

- **Stirring:** a solid dissolves faster in a solvent if it is stirred.
- **Adding more solvent/water:** a solid dissolves faster if there is more solvent.
- **Heating:** a solid dissolves faster in a solvent if the temperature is increased.
- **Making the solid particles smaller:** a solid dissolves faster in a solvent if it is crushed into a fine powder.

These are illustrated in Figure 7.3.

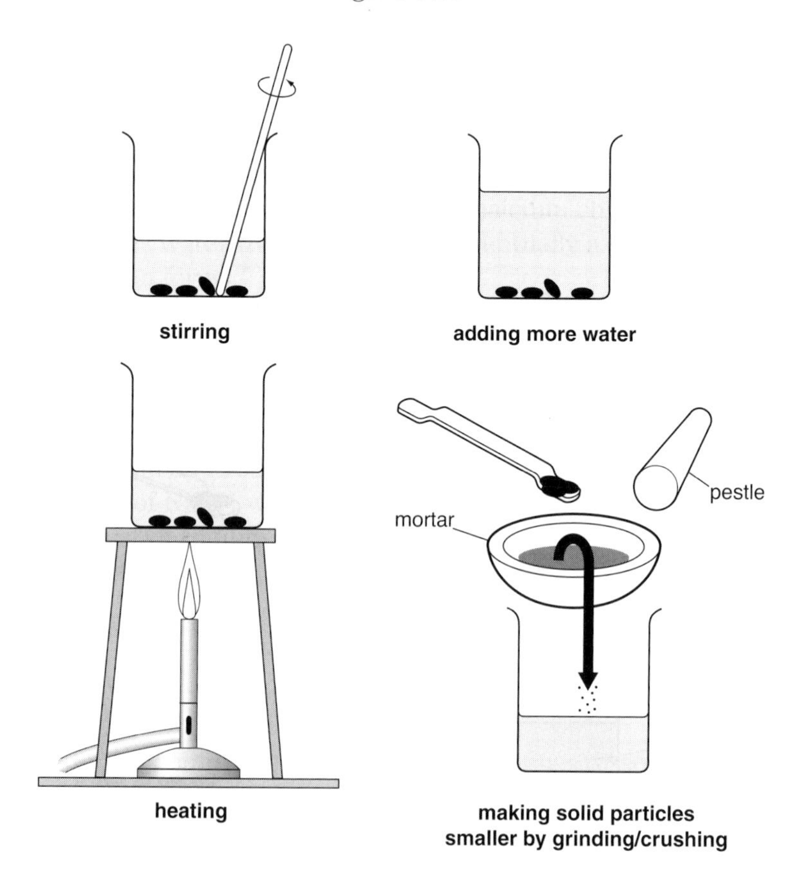

Figure 7.3 Ways of speeding up dissolving

Solubility

Solubility is the mass of a solute that will saturate 100 g of water (the solvent) at a particular temperature. The units of solubility are **g/100 g** of water.

Another way of saying this is that the solubility is the maximum mass of a solute that can dissolve in 100 g of water at a particular temperature.

Typical question

1 What is meant by the term solubility? *[4]*

Answer

1 *The mass [1] of solute that saturates [1] 100 g of water [1] at a particular temperature [1].*

HINT: The most common mistake in this question is to miss the term 'mass' and instead use 'amount', which is incorrect. This is a very common question and you should learn the definition of solubility thoroughly.

Some solubility values

- **Potassium chlorate(V), $KClO_3$** has a solubility value of 14 g/100 g water at 40 °C.

NOTE: This means that 14 g of potassium chloride are required to saturate 100 g of water at 40 °C.

- **Potassium chloride, KCl** has a solubility value of 40 g/100 g water at 40 °C.

More on units of solubility

The units of solubility are important when you are working with graphs and for understanding calculations.

- Mass is measured in g (grams).
  Volume is measured in cm^3 (cubic cm).
  1 cm^3 = 1 ml (millilitre).
- Water has a density of 1 g/cm^3.
  1 cm^3 of water has a mass of 1 g.
  1 g of water has a volume of 1 cm^3.
  100 g of water is the same as 100 cm^3 of water.
- The values for the solubility of most **gases** in water are low, so the units are usually mg/l (milligrams per litre).
  1 mg is $\frac{1}{1000}$th of a gram.
  1 litre is 1000 ml = 1000 cm^3

Preparing a salt

Once a solution of a salt is prepared, it has to be **crystallised** (see page 49). This process involves heating to evaporate some water (usually until half the volume of water is removed). The solution is left to cool and crystallise. Crystallisation occurs because as the solution cools, the solubility of the salt *decreases*.

Solubility of solids

Table 7.2 shows the changes in solubility of potassium chloride (a typical solid) with temperature. As the temperature increases, the solubility of potassium chloride increases.

Table 7.2 The solubility of potassium chloride at different temperatures

Temperature (°C)	0	10	20	30	40	50	60	70	80	90	100
Solubility (g/100 g water)	28	31	34.5	37.5	40	43	45.5	48.5	51	54	56.5

Solubility of gases

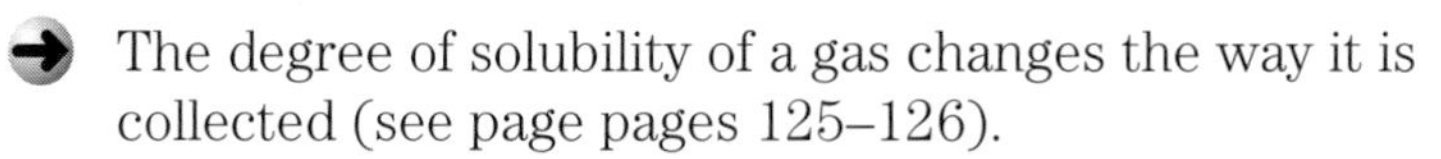

The degree of solubility of a gas changes the way it is collected (see page pages 125–126).

- Oxygen, carbon dioxide and hydrogen are not very soluble in water. These gases can be collected over water.
- Hydrogen chloride, sulphur dioxide and ammonia are very soluble in water. These gases cannot be collected over water.

Thermal pollution

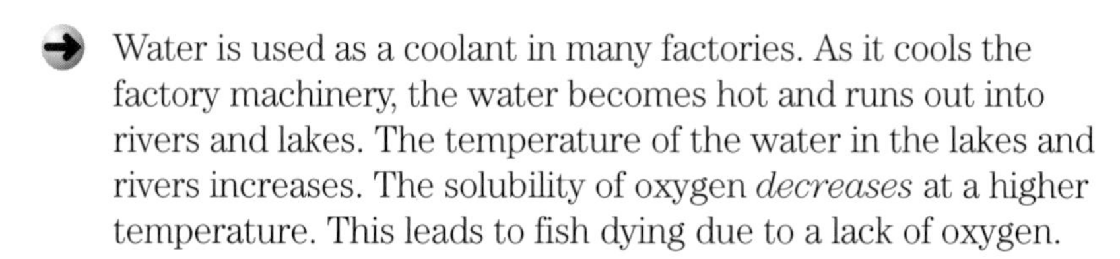

Water is used as a coolant in many factories. As it cools the factory machinery, the water becomes hot and runs out into rivers and lakes. The temperature of the water in the lakes and rivers increases. The solubility of oxygen *decreases* at a higher temperature. This leads to fish dying due to a lack of oxygen.

Typical question

2 Many dead fish were found in a river into which a power station released large amounts of hot water. Explain this. *[3]*

Answer

2 *Solubility of oxygen [1] decreases with temperature [1] so fish die from lack of oxygen [1].*

HINT: Many answers to this type of question miss the whole point which is to do with the solubility of gases in water at different temperatures. Remember that the solubility of a gas *decreases* as temperature *increases*.

Trends in solubility

- **Solids:** as temperature increases, the solubility of solids increases.
- **Gases:** as temperature increases, the solubility of gases decreases.

Practical method of determining the solubility of a solid

The following steps are shown in Figure 7.4.

1 Wear safety glasses. Weigh a clean, dry boiling tube.
2 Add some solid and re-weigh the boiling tube.
3 Add 10 cm^3 of deionised water from a pipette to the boiling tube. A pipette is an accurate method of measuring a volume of a liquid or solution. A liquid is drawn up (sucked) into a pipette using a pipette filler.
4 Place a thermometer in the boiling tube.
5 Heat the boiling tube in a water bath until all the crystals dissolve. Stir gently using the thermometer.
6 Allow the solution to cool slowly and record the temperature when crystals appear.

Figure 7.4 Method of determining the solubility of a solid

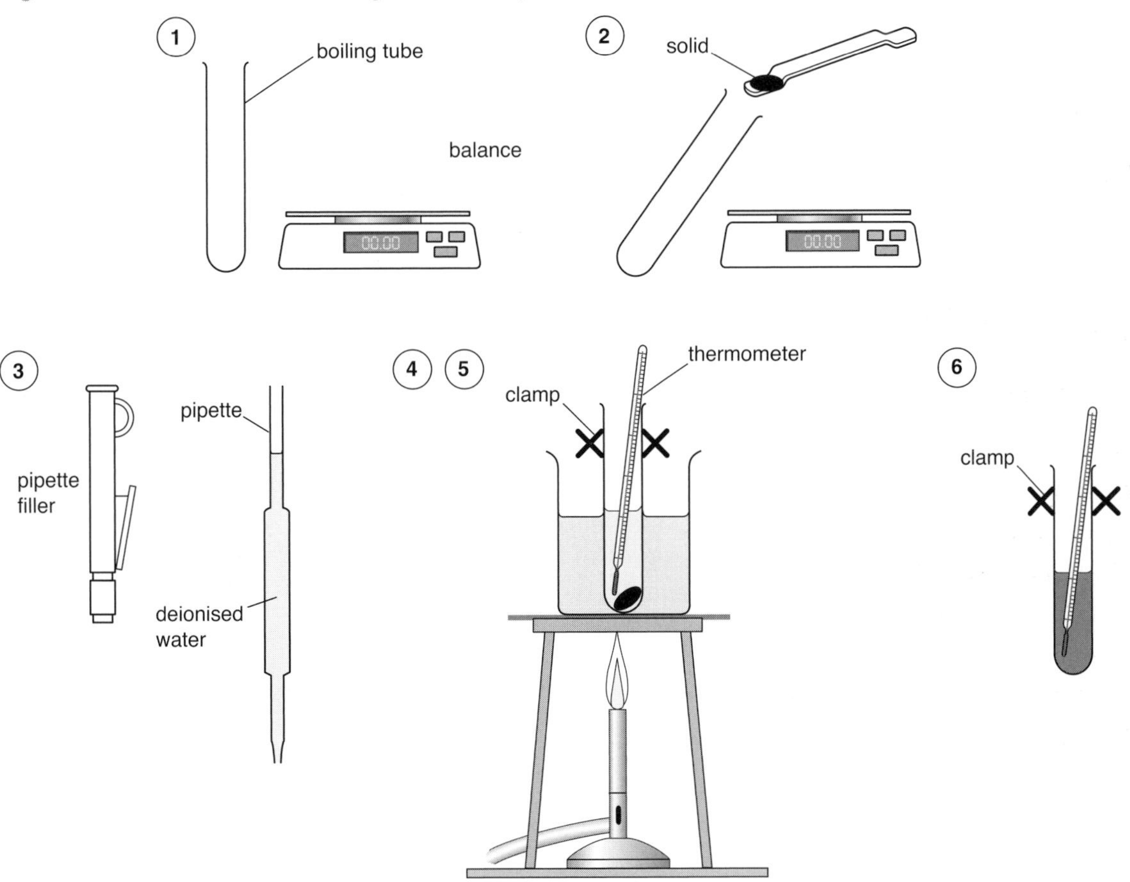

Water of crystallisation

- **Water of crystallisation** is water that is **chemically bonded** into a **crystal structure**.
- **Hydrated** means that solid crystals contain water of crystallisation.
- **Dehydration** means removal of water of crystallisation.
- **Anhydrous** means a powdered solid without water of crystallisation.

Common hydrated substances

Table 7.3 gives the formulae and appearance of copper(II) sulphate, cobalt(II) chloride and sodium carbonate in their hydrated and anhydrous states.

Table 7.3

Salt	Copper(II) sulphate	Cobalt(II) chloride	Sodium carbonate
formula of hydrated salt	$CuSO_4.5H_2O$	$CoCl_2.6H_2O$	$Na_2CO_3.10H_2O$
appearance of hydrated salt	blue crystals	pink crystals	translucent crystals
formula of anhydrous salt	$CuSO_4$	$CoCl_2$	Na_2CO_3
appearance of anhydrous salt	white powder	pale blue powder	white powder

NOTE: In a question asking for the appearance of the salt in its hydrated or anhydrous state, 1 mark is awarded for the correct colour and 1 mark for correctly stating whether the substance is in its crystalline or powdered form.

Degree of hydration

The **degree of hydration** of a substance is the number of moles of water of crystallisation chemically bonded to 1 mole of the compound.

For example, the degree of hydration of $CuSO_4.5H_2O$ is 5. The degree of hydration of a hydrated substance may be determined by heating it to a constant mass, and taking mass measurements before and after heating, or by titration (page 86).

Heating hydrated crystals

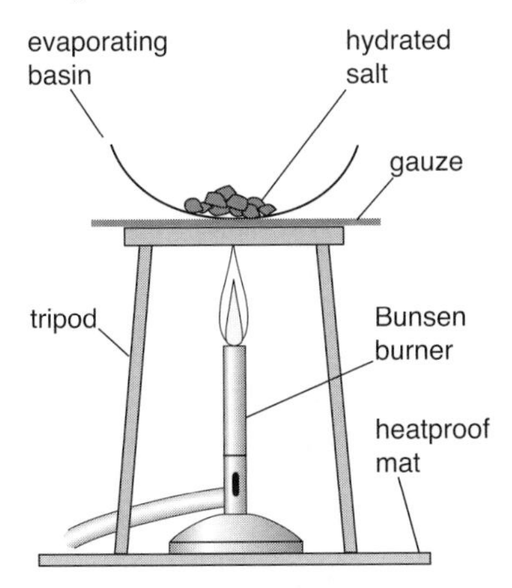

Crystalline hydrated crystals

Heating a hydrated substance changes it to an anhydrous substance and causes a loss in mass (Figure 7.5). On heating, the crystal structure breaks up to leave a powder.

Powdered anhydrous substances

Removing water of crystallisation from a substance is called **dehydration**. Dehydration can be achieved by heating or by chemical means. For example, concentrated sulphuric acid chemically dehydrates hydrated salts (see page 140).

Figure 7.5 Heating hydrated crystals

Drawing and using solubility graphs

Solubility values can be plotted against temperature on a graph and a curve (or less often a line) drawn. The y axis is solubility (g/100 g water). The x axis is temperature (°C).

Example 1

Draw a solubility curve for potassium chlorate(V), $KClO_3$, using the data shown in the table.

Temperature (°C)	0	10	20	30	40	50	60	70	80	90	100
Solubility of $KClO_3$ (g/100 g water)	3	5	7.5	10.5	14	19	24	30	38	46	54

Plot the values shown in the table above on a graph (Figure 7.6).

Figure 7.6

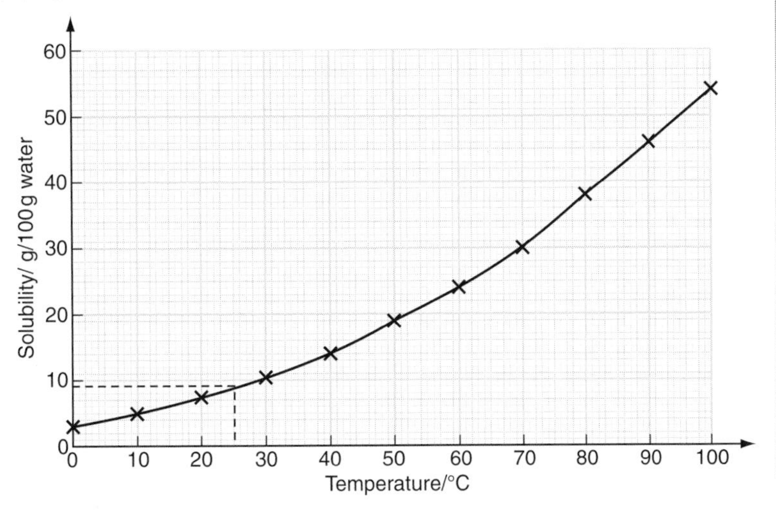

NOTES:

- Plot points using an '×', with the centre of the '×' at the exact coordinate.
- The solubility at 0 °C is not 0 g/100 g water so do not plot (0,0) on a solubility curve.
- Draw the best-fit curve or line on the graph through the points. A curve is usually more suitable than a line.
- The graph allows you to determine solubility values for temperatures in between the given values.

Determining solubility at intermediate values

Example 2

From the solubility graph in Figure 7.6, what is the solubility of potassium chlorate(V) at 25 °C?

On the graph, a vertical line is drawn from the required temperature up to the solubility curve. A second line is drawn horizontally to the solubility axis to determine the solubility at this temperature.

The solubility at 25 °C is 9 g/100 g water.

Determining the mass of crystals formed on cooling a saturated solution

When a **saturated solution** cools down, the solubility of the solid *decreases* as the temperatue *decreases*. The mass of solid that can be dissolved in the solution decreases as the temperature decreases.

The difference between the solubility values is the mass of solid formed on cooling a saturated solution containing 100 g of water.

Example 3

Using the solubility graph in Figure 7.6 on page 69, determine the mass of solid formed on cooling the solution from 75 °C to 25 °C.

The solubility of potassium chlorate(V) at 75 °C is 34 g/100 g water, and its solubility at 25 °C is 9 g/100 g water.

When a **saturated solution** containing 100 g of water is cooled from 75 °C to 25 °C the mass of solid which crystallises is $34 - 9 = 25$ g.

NOTE: Determining the mass of solid which crystallises when the mass of water is *not* 100 g in the saturated solution is a common question (see Example 4).

Example 4

If a **saturated solution** of potassium chlorate(V) containing 50 g of water is cooled from 75 °C to 25 °C, what is the mass of solid crystals formed?

Solubility at 75 °C = 34 g/100 g water
Solubility at 25 °C = 9 g/100 g water

When a saturated solution containing 100 g of water is cooled from 75 °C to 25 °C, the mass of solid that crystallises is 34 − 9 = 25 g.

When a saturated solution containing 50 g of water is cooled from 75 °C to 25 °C, the mass of solid that crystallises is $\frac{25}{2} = 12.5$ g.

Typical questions

The solubility values of potassium nitrate are given in the table below.

Temperature (°C)	0	10	20	30	40	50	60
Solubility of potassium nitrate (g/100 g water)	28	30.5	33	36	39	42	45

1 Determine the mass of potassium nitrate that would be obtained on cooling a saturated solution containing 1000 g of water from 60 °C to 20 °C. *[4]*

2 A solution was prepared by mixing 15 g of potassium nitrate with 50 g of water at 30 °C. Determine whether this solution is saturated or not. *[1]*

Answers

1 Solubility at 60 °C = 45 g/100 g water.
Solubility at 20 °C = 33 g/100 g water.
For a solution with 100 g of water: 45 − 33 *[1]* = 12 g *[1]*
For a solution with 1000 g of water: 12 × 10 *[1]* = 120 g *[1]*

HINT: The most common mistake in this type of question is to leave out the last step where you convert to the mass of water given in the question. *Always* check the question to make sure you have taken this into account. The marks are usually for subtraction of the solubility values at the two temperatures and then either division or multiplication, depending on the mass of water present. (*Don't* subtract the temperature values.)

2 Solubility of potassium nitrate at 30 °C = 36 g/100 g water.
Maximum mass of potassium nitrate which can dissolve in 50 g of water = $\frac{36}{2} = 18$ g.
So a solution containing 15 g of potassium nitrate in 50 g of water is not saturated. *[1]*

(3 g more of the potassium nitrate could be dissolved to make the solution saturated.)

Revision questions

1 What is observed when a freshly cut piece of sodium is left exposed to moist air? *[3]*

2 What is meant by the term 'hard water'? *[2]*

3 Name **one** chemical that causes temporary hardness in water. *[1]*

4 What is the chemical name for washing soda? *[2]*

5 State how the solubility of a gas changes as temperature is increased. *[1]*

6 What is meant by the following terms?
- **a** solvent
- **b** solute
- **c** solution
- **d** saturated solution *[8]*

7 What is water of crystallisation? *[2]*

8 What colour are the following hydrated salts?
- **a** hydrated copper(II) sulphate *[1]*
- **b** hydrated cobalt chloride *[1]*

9 State a method you could use to dehydrate copper(II) sulphate. *[1]*

10 When sodium hydroxide pellets are left exposed to moist air, a white crust forms after several days. What is the chemical name of the white crust? Write a balanced symbol equation for its formation. *[4]*

11 What are the units of solubility? *[1]*

12 State **one environmental** problem caused by excess nitrates in river water. *[1]*

13 Explain the following:
- **a** how temporary hardness arises in water *[4]*
- **b** how washing soda removes hardness from water. *[4]*

14 Four samples of water, labelled A to D, were tested for hardness by shaking with soap solution in a test tube. The height of the lather was measured. A fresh sample of each water was boiled and then retested with soap solution. The height of the lather was again measured. The results are recorded in the following table.

Sample	Height of lather with initial sample + soap	Height of lather with boiled sample + soap
A	15	15
B	1	15
C	1	1
D	1	7

- **a** Which of the samples contained soft water? *[1]*
- **b** Which sample contained only permanent hardness? *[1]*
- **c** Which sample contained only temporary hardness? *[1]*
- **d** Explain your answer to part **c**. *[2]*

15 The solubility curve of potassium nitrate is given below.

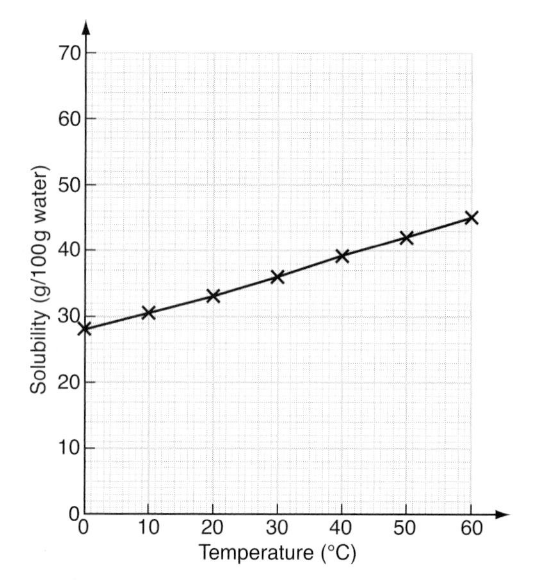

What mass of potassium nitrate would crystallise from solution when a saturated solution containing 50 g of water is cooled from 45 °C to 25 °C? *[4]*

8

Quantitative chemistry

The atom

The **mass** of an atom is largely centred in the nucleus. The **relative atomic mass** (**RAM**) of an atom is measured relative to the mass of an atom of carbon-12 (see page 19). The **relative formula mass** (**RFM**) is the total of the RAMs of *all* the atoms present in the formula of a substance.

The mole

A **mole** of a substance is the standard measurement of amount. One mole of any substance contains 6×10^{23} particles. The number of moles of a substance can be determined in various ways, depending on whether it is a solid, a solution or a gas.

- If you are given a mass of a substance, you need to divide the mass in grams by the RFM.
- If you are given a gas volume in cm^3, then you need to divide the gas volume by 24 000. If the gas volume is given in dm^3, then the gas volume should be divided by 24 to calculate the number of moles.
- If a solution volume is given in cm^3 and a concentration in mol/dm^3, the moles can be determined by multiplying the two values and dividing by 1000.

The following expressions can be used to calculate the number of moles of a solid, a solution and a gas:

solids: $$\text{moles} = \frac{\text{mass (g)}}{\text{RFM}}$$

gases: $$\text{moles} = \frac{\text{gas volume (cm}^3\text{)}}{24\,000}$$

solutions: $$\text{moles} = \frac{\text{solution volume (cm}^3\text{)} \times \text{concentration (mol/dm}^3\text{)}}{1000}$$

Percentage composition

➜ From the formula of a compound, we can calculate the **percentage by mass** of each of the elements within the compound. This shows how we calculate the percentage composition by mass:

% of element M in a compound

$$= \frac{\textbf{number of atoms of M in compound} \times \textbf{RAM (M)}}{\textbf{RFM of compound}}$$

Example 1

Calculate the percentage composition of carbon in ethane, C_2H_6, and ethene, C_2H_4.

	Ethane	Ethene
Formula:	C_2H_6	C_2H_4
RFM:	30	28
Mass of C:	24	24
% C:	$\frac{24}{30} \times 100 = 80\%$	$\frac{24}{28} \times 100 = 85.71\%$

NOTE: You may be asked to calculate the percentage of water of crystallisation rather than of an element within a compound, but just consider water to be one unit with an RFM of 18. Remember to include the degree of hydration. If there are five water molecules as given in the example below, then multiply 18 by 5 to get the total mass of water in the compound.

Example 2

Find the percentage water by mass in copper(II) sulphate-5-water.

Formula:	$CuSO_4.5H_2O$
RFM:	$64 + 32 + (4 \times 16) + 5 \times (2 + 16) = 250$
Mass of H_2O:	$5 \times 18 = 90$
% H_2O:	$\frac{90}{250} \times 100 = 36\%$

Using percentage composition to find formulae

➜ If the percentage composition of a compound is given, the **simplest formula** of the compound can be determined. This is done by using percentage values as mass values. The mass values are converted to moles by dividing by the RFM. The moles are converted to a simple ratio – this is best achieved by making the lowest mole value equal to 1, and then dividing through the other mole values by the lowest mole value.

The ratio is the simplest ratio of the elements (and water of crystallisation if present) in the compound. A formula written from the simplest ratio is called the **empirical formula**.

Example 1

An oxide of copper contains 80% copper and 20% oxygen by mass. Find the formula of the compound.

	% by mass	RFM	% by mass ÷ RFM (= mole ratio)	Simple ratio
copper	80	64	$\frac{80}{64} = 1.25$	1
oxygen	20	16	$\frac{20}{16} = 1.25$	1
Empirical formula = CuO				

In this example the mole values are the same, so the simplest ratio is 1:1.

Example 2

Hydrated barium chloride contains 56.2% barium, 29.0% chlorine and 14.8% water by mass. Find the formula of the hydrated barium chloride.

	% by mass	RFM	% by mass ÷ RFM (= mole ratio)	Simple ratio
barium	56.2	137	$\frac{56.2}{137} = 0.41$	1
chlorine	29.0	35.5	$\frac{29.0}{35.5} = 0.82$	2
water	14.8	18	$\frac{14.8}{18} = 0.82$	2
Empirical formula = $BaCl_2.2H_2O$				

In this example the smallest mole number is 0.41.
This is made equal to 1 and divided into the other moles.
The others work out at 2.

Using mass values to determine formulae

Example 1

Given that 1.06 g of magnesium combines with oxygen to give 1.77 g of magnesium oxide, find the formula of the oxide of magnesium.

This is done practically by heating a certain mass of magnesium in a crucible with a lid, which is raised periodically to let fresh air in. The magnesium is heated to a constant mass to ensure that all the magnesium has combined to form the oxide. The following method is followed.

1 Find the mass of the empty crucible: 16.18 g (1)
2 Find the mass of the crucible and some magnesium: 17.24 g (2)
3 Mass of magnesium
$= (2) - (1) = 17.24 - 16.18 = 1.06$ g (3)
4 Find the mass of the crucible after heating to a constant mass: 17.95 g (4)
5 Mass of oxygen combined
$= (4) - (2) = 17.95 - 17.24 = 0.71$ g

Using the information obtained above we can now calculate the formula of the oxide of magnesium.

Element	magnesium	oxygen
Mass (g)	1.06	0.71
RFM	24	16
Moles	$\frac{1.06}{24} = 0.044$	$\frac{0.71}{16} = 0.044$
Ratio	1	1
Empirical formula	**MgO**	

NOTE: 16 is used for the RFM of oxygen as it is oxygen *atoms* combined in the formula. Because the ratio is found to be 1 Mg atom to 1 O atom, we say that the **empirical formula** is MgO, but it could also be Mg_2O_2 or Mg_3O_3, etc. as the ratio in these compounds is the same.

Example 2

Given that 4.0 g of hydrated copper(II) sulphate, $CuSO_4.nH_2O$, produces 2.56 g of the anhydrous copper(II) sulphate, $CuSO_4$, on heating to constant mass, find the value of n in the formula of the hydrated salt.

Mass of hydrated salt: 4.0 g
Mass of anhydrous salt: 2.56 g
Mass of water lost = 4.0 − 2.56 = 1.44 g

Compound	copper(II) sulphate	water
Formula	$CuSO_4$	H_2O
Mass (g)	2.56 g	1.44 g
RFM	160	18
Moles	$\frac{2.56}{160} = 0.0160$	$\frac{1.44}{18} = 0.080$
Ratio (÷0.0160)	1	5
Empirical formula	**$CuSO_4.5H_2O$**	

You can see from the empirical formula that the value of n is 5.

Using balanced symbol equations quantitatively

A balanced symbol equation such as the one below can be read **quantitatively**.

$$2Pb(NO_3)_2 \xrightarrow{\text{heat}} 2PbO + 4NO_2 + O_2$$

This equation shows that 2 moles of $Pb(NO_3)_2$, when heated to constant mass, break down to produce 2 moles of PbO, 4 moles of NO_2 and 1 mole of O_2.

The $Pb(NO_3)_2$ is heated to constant mass to ensure that it all decomposes.

If there is a different number of moles of $Pb(NO_3)_2$ to start with, the balancing number in the equation still gives the ratio of how many moles react or are produced.

There are three steps to follow.

1. Using the mass of one of the reactants, which will be given to you, calculate the number of moles of this substance.
2. Using the balancing numbers in the equation, calculate the number of moles of the substance asked in the question.
3. Change the number of moles of this substance to mass or volume as required.

Example 1

$Pb(NO_3)_2$ undergoes thermal decomposition according to the equation:

$$2Pb(NO_3)_2 \rightarrow 2PbO + 4NO_2 + O_2$$

3.31 g of $Pb(NO_3)_2$ were heated to constant mass. Calculate the mass of PbO formed.

Method 1

1 RFM of $Pb(NO_3)_2 = 207 + 2 \times (14 + 3 \times 16) = 331$

As it is a solid, moles $= \frac{\text{mass}}{\text{RFM}} = \frac{3.31}{331} = 0.01$ moles of $Pb(NO_3)_2$.

2 In the balanced symbol equation, 2 moles of $Pb(NO_3)_2$ form 2 moles of PbO.
So 0.01 moles of $Pb(NO_3)_2$ forms 0.01 moles of PbO.

3 0.01 moles of PbO can be converted to mass by multiplying by its RFM.
RFM of PbO = 207 + 16 = 223
Mass of PbO formed = 0.01 × 223 = 2.23 g

Method 2

This type of calculation can be set out in a table below the balanced symbol equation.

	$2Pb(NO_3)_2$	→	2PbO	+	$4NO_2$	+	O_2
mass	3.31		**2.23 g				
RFM	331		**223				
moles	0.01		*0.01				

1 Put in the mass you have been given and calculate the RFM value of that substance. Divide the mass by the RFM to calculate the number of moles. This is shown in the $Pb(NO_3)_2$ column.

2 Then calculate the other moles using the balancing numbers. 0.01 moles of $Pb(NO_3)_2$ produces 0.01 moles of PbO (* in the table).

3 Calculate the RFM of PbO and multiply it by the number of moles to determine the mass of PbO (** in the table).

HINT: The types of calculations shown in Example 1 are often asked in questions. The most common mistake is to calculate the RFM and to multiply it by the balancing number before calculating the number of moles. *Remember that the RFM is for one formula.* The balancing numbers are for that particular equation. In Example 1, the error would be to use '662' as the RFM of $Pb(NO_3)_2$ because it has a '2' in front of it. Remember when using the table method to *work down* to moles, *go across* using the balancing numbers and *work up* to mass.

Example 2

27 kg of aluminium were heated in a stream of oxygen until constant mass was achieved. Determine the volume of oxygen gas required to react and the mass of aluminium oxide formed.

$$4Al + 3O_2 \rightarrow 2Al_2O_3$$

Method 1

27 kg of aluminium is 27 000 g.

As it is a solid, moles $= \frac{\text{mass}}{\text{RFM}} = \frac{27\,000}{27} = 1000$ moles of Al.

In the balanced symbol equation, 4 moles of Al react with 3 moles of O_2 to form 2 moles of Al_2O_3.

So 1000 moles of Al react with $\frac{1000}{4} \times 3 = 750$ moles of O_2

to form $\frac{1000}{2} = 500$ moles of Al_2O_3.

750 moles of O_2 can be converted to gas volume by multiplying by 24 dm^3.

$750 \times 24 = 18\,000$ dm^3 of oxygen gas required.

RFM of $Al_2O_3 = 2 \times 27 + 3 \times 16 = 102$.

Mass of Al_2O_3 formed $= 500 \times 102 = 51\,000$ g $= 51$ kg.

Method 2

The table shows step 1 in the first column, step 2 working out other moles* and step 3 working out gas volume and mass**.

	4Al	+	**$3O_2$**	→	**$2Al_2O_3$**
mass	27 000				**51 000
RFM	27				**102
moles	1000		*750		*500
gas volume			**18 000		

HINT: Questions are set using 'kg' to ensure you know to use mass in grams. *Always* convert to grams before working out moles.

Working with solutions

➜ Solutions are slightly more complicated as the number of moles depends on the **volume** of the solution (in cm^3) and on the **concentration** of the solution (in mol/dm^3).

A solution that has a concentration of 1 mol/dm^3 has 1 mole of the solute dissolved in 1 dm^3 of solution. If you had 500 cm^3 of a solution of concentration 1 mol/dm^3, you would have 0.5 moles of the solute in that volume.

1 dm^3 is the same as 1 litre.
1 cm^3 is the same as 1 ml.
There are 1000 cm^3 in 1 dm^3 (1000 ml in 1 litre).

Solution calculations

The **number of moles** in a solution may be calculated using this equation (see also page 73):

$$\textbf{moles} = \frac{\textbf{solution volume (cm}^3\textbf{)} \times \textbf{concentration (mol/dm}^3\textbf{)}}{\textbf{1000}}$$

Making a solution

When making a solution a certain mass of a solid is dissolved in a certain volume of water. The **concentration** is found using the above equation.

Example 1

1.4 g of KOH (potassium hydroxide) were dissolved completely in 100 cm^3 of water. Calculate the concentration of the solution formed.

$$1.4 \text{ g of solid KOH} = \frac{1.4}{56} = 0.025 \text{ moles.}$$

0.025 moles of KOH are present in 100 cm^3.

For a solution:

$$\text{moles} = \frac{\text{solution volume (cm}^3) \times \text{concentration (mol/dm}^3)}{1000}$$

$$0.025 = \frac{100 \times \text{concentration}}{1000}$$

$$\text{concentration} = \frac{0.025 \times 1000}{100} = 0.25 \text{ mol/dm}^3$$

More simply, this is 10 × the number of moles in 100 cm^3.

Determining moles of solute in a solution

➜ The same equation is also used to calculate the number of moles of a **solute** in a certain volume of solution.

Example 2

45.0 cm^3 of a 0.1 mol/dm^3 solution of hydrochloric acid were used in a titration. Calculate the number of moles of hydrochloric acid used.

For a solution:

$$\text{moles} = \frac{\text{solution volume (cm}^3) \times \text{concentration (mol/dm}^3)}{1000}$$

$$\text{moles} = \frac{45.0 \times 0.1}{1000} = 0.0045 \text{ moles of HCl}$$

Dilution

➜ When a solution is **diluted** (by adding water) the concentration changes, but the total number of moles of solute does not. This is because water has been added but no more solute has been added.

Example 3

25.0 cm^3 of a 0.2 mol/dm^3 solution of sodium hydroxide is diluted to 250.0 cm^3 in a volumetric flask. Calculate the new concentration of the solution.

This is a 1 in 10 dilution (25.0 cm^3 up to 250.0 cm^3) so the concentration of the sodium hydroxide has been reduced by a factor of 10. The new concentration is 0.02 mol/dm^3.

Titration

A **titration** is a method of reacting two solutions together to determine the number of moles of one of the solutes (in one of the solutions) and then usually to work out an unknown concentration.

Method of titration

Titration is also covered on page 48.

1 A measured volume of one solution is placed in a conical flask (using a pipette).
2 An indicator is added.
3 A burette is charged (filled) with the other solution.
4 The second solution is run out of the burette into the conical flask while swirling the solution in the flask. The second solution is added until the indicator just changes colour.
5 The volume of solution added from the burette is recorded. This volume is called a **titre** (volume of titrated solution added).
6 The titration is carried out three times. The first value is a rough value and should be ignored in calculations. The second and third values are accurate and should be averaged to find the average titration volume (the **average titre**).

Titration calculations

There are three steps to follow in titration calculations.

1 Enough information will be given to determine the moles of one solute.
2 The balanced symbol equation between the two solutes is used to calculate the other number of moles.
3 The concentration of the other solution can be determined from the number of moles and the solution volume.

Typical question

1 25.0 cm^3 of a solution of sulphuric acid of unknown concentration was placed in a conical flask using a pipette. Phenolphthalein indicator was added and the solution was titrated against 0.1 mol/dm^3 sodium hydroxide solution. The average titre was found to be 17.5 cm^3.

$$2NaOH + H_2SO_4 \rightarrow Na_2SO_4 + 2H_2O$$

a Calculate the number of moles of sodium hydroxide used in this titration. *[2]*

b Calculate the number of moles of sulphuric acid which will react with this number of moles of sodium hydroxide. *[2]*

c Calculate the concentration of the sulphuric acid in mol/dm^3. *[2]*

Answer

1 a moles of NaOH $= \dfrac{\text{solution volume (cm}^3) \times \text{concentration (mol/dm}^3)}{1000}$

$= \dfrac{17.5 \times 0.1}{1000} = 0.00175$ [2]

b The moles of H_2SO_4 is half the number of moles of NaOH, as 2 moles of NaOH reacts with 1 mole of H_2SO_4 in the equation.

moles of $H_2SO_4 = \dfrac{0.00175}{2} = 0.000875$ [2]

c concentration of $H_2SO_4 = \dfrac{\text{moles} \times 1000}{\text{solution volume (cm}^3)}$

$= \dfrac{0.000875 \times 1000}{25.0} = 0.035\ \text{mol/dm}^3$ [2]

HINT: As you can see, these questions are usually structured but again occur in three stages. First, calculate the number of moles of one substance, then determine the number of moles of the other (using the balanced symbol equation) and thirdly calculate a quantity such as concentration for the other substance.

Sometimes the calculation gets a little more complicated but the method is the same. A dilution may be added to one of the solutions.

Typical question

2 A solution of ethanoic acid (CH_3COOH) is diluted by placing 10.0 cm^3 of the solution into a 250.0 cm^3 volumetric flask and making up the volume using deionised water. 25.0 cm^3 of this diluted solution were placed in a conical flask and titrated against 0.2 mol/dm^3 potassium hydroxide (KOH) solution, using phenolphthalein indicator. 15.7 cm^3 of the potassium hydroxide solution were required for neutralisation.

$$CH_3COOH + KOH \rightarrow CH_3COOK + H_2O$$

a Calculate the number of moles of KOH used in this titration. [2]

b Calculate the number of moles of CH_3COOH which reacted with this number of moles of KOH. [2]

c Calculate the concentration of the diluted CH_3COOH solution. [2]

d Calculate the concentration of the undiluted CH_3COOH solution. [2]

Answer

2 a moles of KOH $= \dfrac{15.7 \times 0.2}{1000} = 0.00314$ [2]

b moles of $CH_3COOH = 0.00314$ (1:1 ratio in balanced symbol equation) [2]

c concentration of diluted CH_3COOH solution $= \dfrac{0.00314 \times 1000}{25} = 0.1256\ \text{mol/dm}^3$ [2]

d concentration of undiluted CH_3COOH solution $= 0.1256 \times 25 = 3.14\ \text{mol/dm}^3$.
(The dilution factor was 25, as 10 cm^3 was diluted to 250 cm^3.) [2]

Finding a formula from titration values

This type of question is usually to find the number of moles of water of crystallisation in a hydrated salt. The mole answers can be compared as before to get the simplest ratio of the salt to the water of crystallisation.

Typical question

2.86 g of a sample of hydrated sodium carbonate, $Na_2CO_3.xH_2O$ were dissolved in 250.0 cm^3 of water. A 25.0 cm^3 sample of this solution was placed in a conical flask and titrated with 0.1 mol/dm^3 hydrochloric acid using methyl orange indicator. The average titre was 20.0 cm^3.

$$Na_2CO_3 + 2HCl \rightarrow 2NaCl + H_2O + CO_2$$

a Calculate the number of moles of hydrochloric acid used in this titration. *[1]*
b Calculate the number of moles of sodium carbonate in 25.0 cm^3 of the solution. *[1]*
c Calculate the number of moles of sodium carbonate present in 250.0 cm^3 of solution. *[1]*
d Calculate the mass of sodium carbonate, Na_2CO_3, present in the initial sample. *[1]*
e Calculate the mass of water present in the initial sample. *[1]*
f Calculate the number of moles of water in the initial sample. *[1]*
g Using the answers to parts c and f, determine the value of *x* in $Na_2CO_3.xH_2O$. *[1]*

Answer

a moles of HCl $= \frac{20.0 \times 0.1}{1000} = 0.002$ *[1]*

b moles of Na_2CO_3 in 25.0 cm^3 = 0.001 (0.002 ÷ 2 due to ratio in equation) *[1]*

c moles of Na_2CO_3 in 250.0 cm^3 = 0.001 × 10 = 0.01 *[1]*

d mass of Na_2CO_3 in 250.0 cm^3 = 0.01 × 106 = 1.06 g *[1]*

e mass of water = 2.86 − 1.06 = 1.8 g *[1]*

f moles of water $= \frac{1.8}{18} = 0.1$ *[1]*

g moles of Na_2CO_3 = 0.01; moles of water = 0.1
simplest ratio $Na_2CO_3 : H_2O$ = 1:10
so x = 10 *[1]*

Using Avogadro's Law

Avogadro's Law makes gas volume calculations easier. If it is a gas volume you are given and a gas volume that you are asked to find, then the gas volumes are in the same ratio as the balancing numbers in the equation.

Typical question

Ammonia reacts with oxygen according to the equation:

$$4NH_3\ (g) + 5O_2\ (g) \rightarrow 4NO\ (g) + 6H_2O\ (g)$$

Calculate the volume of oxygen required to react completely with 10.0 cm^3 of ammonia, NH_3. *[2]*

Answer

The ratio of ammonia, NH_3, to oxygen, O_2, in the equation is 4:5.

Volume of oxygen $= \frac{10}{4} \times 5 = 12.5\ cm^3$ *[2]*

HINT: Avogadro's Law questions always state 'using Avogadro's Law or otherwise, calculate the volume of ...'. Don't make the question more complicated than it is by trying to calculate moles of gas. You are more likely to make mistakes if you attempt this.

Revision questions

1 To what atom are the masses of all atoms compared? *[2]*

2 State Avogadro's Law. *[3]*

3 Determine the RFM of the following compounds:
a H_2SO_4 **b** $Ca(OH)_2$ **c** $Al_2(SO_4)_3$
d K_2CO_3 **e** $FeCl_3$ *[5]*

4 Calculate the mass of calcium oxide, CaO, which would be produced by heating a sample of 5 g of calcium carbonate, $CaCO_3$, to constant mass.

$$CaCO_3 \rightarrow CaO + CO_2$$ *[5]*

5 What mass of magnesium oxide would be produced when 1.2 g of magnesium powder are burned completely in oxygen?

$$2Mg + O_2 \rightarrow 2MgO$$ *[5]*

6 Aluminium reacts with iron(III) oxide according to the equation:

$$2Al + Fe_2O_3 \rightarrow Al_2O_3 + 2Fe$$

Calculate the mass of iron(III) oxide required to react with 54 kg of aluminium. *[5]*

7 What mass of iron would be produced in the reaction in question **6** above? *[3]*

8 14 cm^3 of oxygen gas react with sulphur dioxide to form sulphur trioxide as part of the industrial manufacture of sulphuric acid.

$$2SO_2 + O_2 \rightarrow 2SO_3$$

Calculate the volume of sulphur dioxide, SO_2, required to react with 14 cm^3 of oxygen using Avogadro's Law or otherwise. *[2]*

9 In what part of the atom is the mass largely centred? *[1]*

10 An oxide of iron contains 27.6% oxygen by mass.

a Calculate the percentage by mass of iron in this oxide. *[1]*

b Determine the empirical formula of the oxide. *[3]*

11 0.6 g of magnesium ribbon reacts completely with 0.1 mol/dm^3 hydrochloric acid according to the equation:

$$Mg + 2HCl \rightarrow MgCl_2 + H_2$$

a Calculate the volume of 0.1 mol/dm^3 hydrochloric acid required to react completely with the magnesium. *[4]*

b Calculate the volume of hydrogen gas produced in this reaction. *[2]*

12 25.0 cm^3 of a solution of potassium hydroxide were placed in a conical flask along with a few drops of phenolphthalein indicator. A burette was filled with 0.25 mol/dm^3 sulphuric acid and the average titration figure was 19.5 cm^3.
The equation for the reaction is:

$$2KOH + H_2SO_4 \rightarrow K_2SO_4 + 2H_2O$$

a Calculate the number of moles of sulphuric acid used. *[2]*

b Calculate the number of moles of potassium hydroxide in 25.0 cm^3 of solution. *[2]*

c Calculate the concentration of the potassium hydroxide solution. *[2]*

13 0.56 g of an unknown metal hydroxide, MOH, was dissolved in 100.0 cm^3 of water. 25.0 cm^3 of the solution were titrated with 0.2 mol/dm^3 hydrochloric acid using a suitable indicator. The volume of acid required for complete neutralisation was found to be 17.5 cm^3. The equation for the reaction can be represented as:

$$MOH + HCl \rightarrow MCl + H_2O$$

a Calculate the number of moles of hydrochloric acid used. *[2]*

b Calculate the number of moles of MOH in 25.0 cm^3 of the solution. *[2]*

c Calculate the number of moles of MOH in 100.0 cm^3 of the solution. *[2]*

d Using the mass of MOH and the number of moles, calculate the RFM of MOH. *[2]*

e Calculate the RAM of M. *[2]*

f Using your *Data Leaflet*, determine the identity of M. *[1]*

14 Explain how you would prepare a burette for use in a titration. *[3]*

15 A sample of hydrated sodium carbonate, $Na_2CO_3.xH_2O$, is heated to constant mass in an evaporating basin. The measurements below are taken at 5-minute intervals.

Mass of evaporating basin	= 122.400 g
Mass of evaporating basin and hydrated sample	= 122.900 g
Mass of evaporating basin and sample after 5 minutes heating	= 122.714 g
Mass of evaporating basin and sample after 10 minutes heating	= 122.612 g
Mass of evaporating basin and sample after 15 minutes heating	= 122.612 g

a Calculate the mass of anhydrous sodium carbonate present at the end of the experiment. *[2]*

b Calculate the number of moles of anhydrous sodium carbonate present at the end of the experiment. *[2]*

c Calculate the mass of water lost by heating. *[2]*

d Calculate the number of moles of water lost by heating. *[2]*

e Using your answer to parts **b** and **d**, determine the value of x in $Na_2CO_3.xH_2O$. *[2]*

9

Rates of reaction

The **rate** (the speed at which the reactants change into products) at which any chemical reaction occurs depends on the following factors:

- surface area/size of solid particles
- concentration of solutions
- temperature
- presence of a catalyst.

Each of these factors can be studied practically. There are different methods of measuring rate. All the methods measure a quantity against time, for example a change in mass or a change in gas volume.

Measuring a change in mass

The reaction between marble chips (calcium carbonate, $CaCO_3$) and hydrochloric acid can be used to investigate how the size of solid particles affects the rate of reaction.

$$CaCO_3\,(s) + 2HCl\,(aq) \rightarrow CaCl_2\,(aq) + H_2O\,(l) + CO_2\,(g)$$

Mass is lost during this reaction as carbon dioxide is escaping from the reaction vessel. Recording the loss in mass over a certain period of time at regular intervals using the apparatus shown in Figure 9.1 gives an indication of the rate of the reaction.

The cotton wool stops any loss of liquid from the flask. There is a large amount of **effervescence** (bubbling) which can cause the chemicals to splash out.

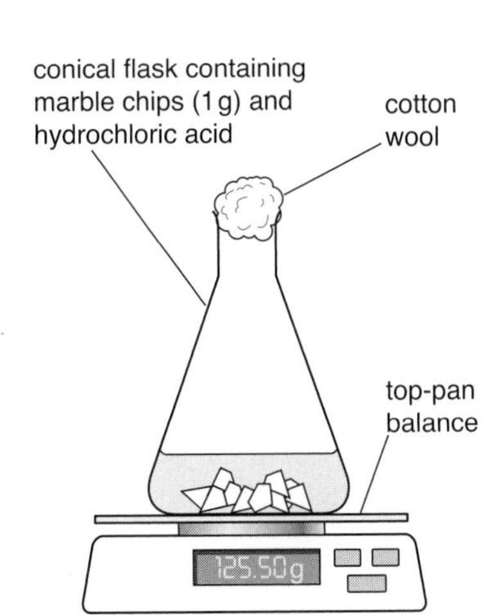

Figure 9.1 Apparatus used to investigate the effect of particle size on rate of reaction

1 g of large marble chips was used and then the experiment was repeated with 1 g of smaller marble chips. The results are shown in the graph of mass against time in Figure 9.2.

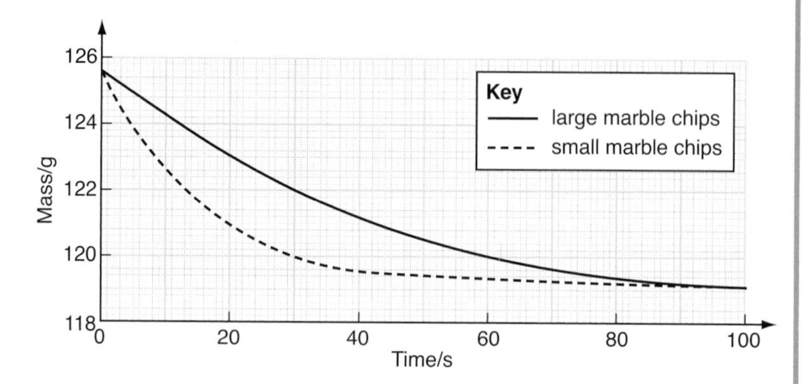

Figure 9.2 Graph to show how particle size affects rate of reaction

From the graph you can see that the reaction occurred more rapidly with smaller marble chips. The smaller marble chips have a much larger surface area which contacts with the acid. This causes an increased rate of reaction.

As the same mass of marble chips and the same volume and concentration of acid were used, both reactions will have the same final loss in mass. This is why the graph starts and levels off at the same mass.

The steeper slope of the curve for small marble chips indicates that the mass is decreasing more quickly. So the **rate of reaction** is higher.

Measuring gas volume

If a reaction produces a gas, the best method to measure the rate of reaction is by measuring gas volume over a period of time. A gas syringe is attached to a sealed reaction vessel (Figure 9.3).

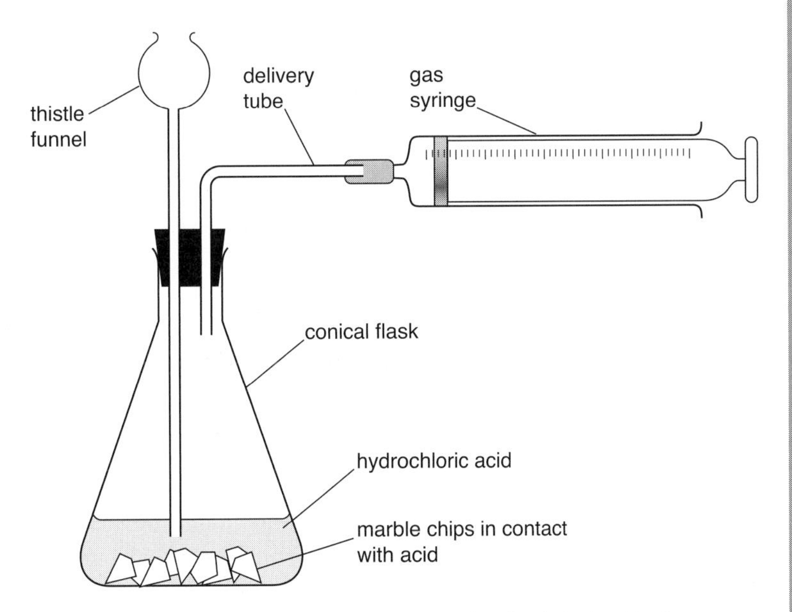

Figure 9.3 Apparatus used to investigate rate of reaction by measuring the volume of gas produced

HINT: The marks for the diagrams are for the labels. Marks are usually divided into three sections.

1. Preparation of gas – 2 marks. You need to draw the reactants in contact with each other and the conical flask with the thistle funnel to the correct depth.
2. Connection – 1 mark. You need to draw a correctly fitted delivery tube.
3. Collection – 2 marks. You need to draw a gas syringe correctly connected to the end of the delivery tube.

A diagram drawn perfectly with no labels is worth 0 marks. The items underlined above are the standard labels required.

Figure 9.3 shows the apparatus used to produce carbon dioxide gas in the reaction between marble chips and hydrochloric acid. The volume of carbon dioxide is measured by taking readings from the gas syringe at various time intervals.
A graph is then plotted of volume of gas produced against time (Figure 9.4, overleaf).

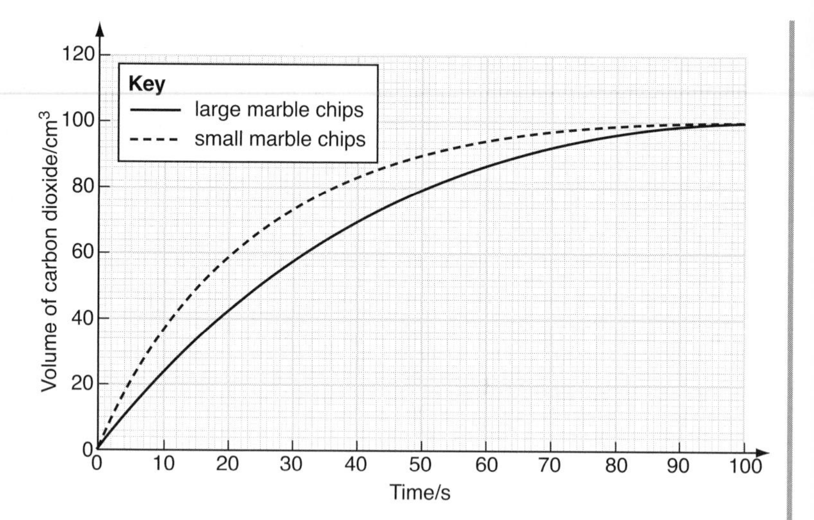

Figure 9.4 Graph to show how particle size affects the volume of gas produced

The volume of gas produced increases more rapidly when small marble chips are used. The lines on the graph start and end at the same gas volume because the same mass of marble chips and the same volume and concentration of hydrochloric acid are used.

Measuring production of a solid precipitate

➜ Sodium thiosulphate solution, $Na_2S_2O_3$ (aq), reacts with dilute hydrochloric acid according to the equation:

$$Na_2S_2O_3\,(aq) + 2HCl\,(aq) \rightarrow 2NaCl\,(aq) + S\,(s) + SO_2\,(g) + H_2O\,(l)$$

This reaction produces solid sulphur. If the solutions are mixed in a conical flask placed on a piece of white paper with a black cross on it, then as the solid sulphur is produced during the reaction it obscures the cross. The time is measured until the cross can no longer be seen when viewed from above. The higher the concentration of sodium thiosulphate solution, the less time is taken for the cross to disappear, which means a higher rate of reaction. The rate equation is:

$$\textbf{rate} = \frac{\textbf{1}}{\textbf{time}}$$

The units of rate are s^{-1}

Explanation in terms of particles

At a higher concentration of solution:

- there are more particles present in the same volume
- which leads to more collisions between the reacting particles
- which leads to more successful collisions
- in a given period of time
- which increases the rate of reaction.

Effect of temperature on rate of reaction

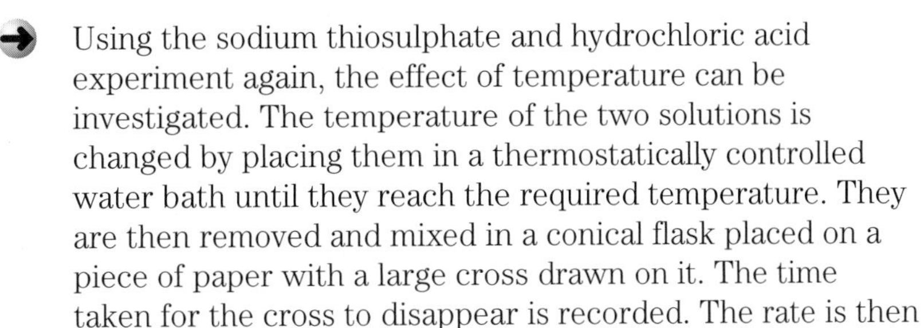

Using the sodium thiosulphate and hydrochloric acid experiment again, the effect of temperature can be investigated. The temperature of the two solutions is changed by placing them in a thermostatically controlled water bath until they reach the required temperature. They are then removed and mixed in a conical flask placed on a piece of paper with a large cross drawn on it. The time taken for the cross to disappear is recorded. The rate is then calculated using the rate equation shown on page 92.

As the temperature of the reactants is increased, the time for the cross to disappear decreases which means there is a higher rate of reaction.

Explanation in terms of particles

At higher temperatures:

- the particles have greater energy and move faster
- which leads to more successful collisions
- in a given period of time
- which increases the rate of reaction.

HINT: The link between particles and rate is important.

- You should always be able to explain why increasing concentration causes an increase in rate and why an increase in temperature causes an increase in rate.
- You should be able to state the names of the particles which have increased in concentration and the particles which are colliding to react (but if you cannot, just write the general statements down using the word 'particle').

Presence of a catalyst

The reaction for the study of the presence of a **catalyst** is the decomposition of hydrogen peroxide using the catalyst **manganese(IV) oxide** (also called manganese dioxide). Oxygen gas is given off.

$$2H_2O_2 \xrightarrow{MnO_2} 2H_2O + O_2$$

This reaction is again monitored using a gas syringe to measure the volume of oxygen gas produced (for the apparatus, see Figure 9.3 on page 91). The results of the investigation are shown in a graph of volume of gas produced against time (Figure 9.5).

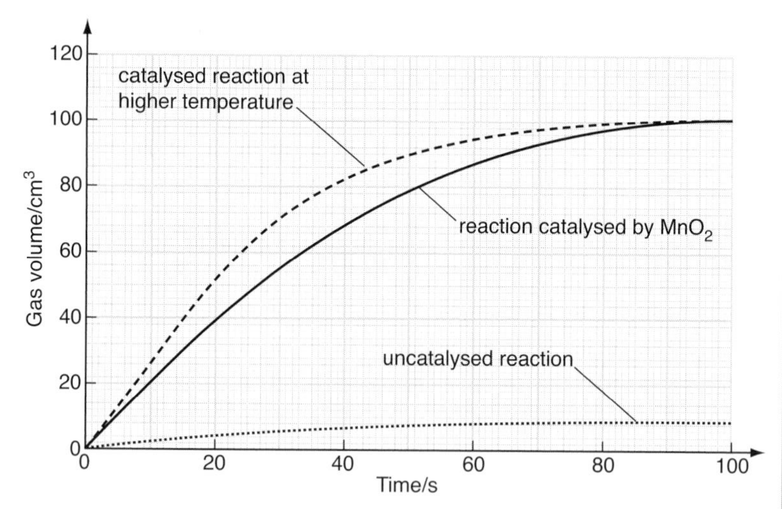

Figure 9.5 Graph of volume of oxygen produced against time in the decomposition reaction of hydrogen peroxide

From the graph in Figure 9.5, the catalysed reaction happens at a higher rate than the uncatalysed reaction, and the catalysed reaction at a higher temperature occurs at the highest rate.

The following four points are the ones to note for the graph at higher temperature when all other factors are kept the same (i.e. same concentration, volume, etc.):

1. gas volume starts at zero
2. gas volume is higher at each time
3. graph levels off earlier
4. but ends at same final gas volume.

HINT: Learn the four points above. They are important as they form the basis of many questions for gas volume as a measure of rate of reaction. At a lower temperature the gas volume would start at zero, stay lower, level off later but end at the same final gas volume.

Metal + acid reactions

Both temperature and concentration can be studied using the following two reactions:

zinc and hydrochloric acid $Zn + 2HCl \rightarrow ZnCl_2 + H_2$

magnesium and hydrochloric acid $Mg + 2HCl \rightarrow MgCl_2 + H_2$

Both reactions produce hydrogen gas, so gas volume is used to measure the rate of the reactions.

- A change in temperature will alter the shape of the gas volume graph – it doesn't make any difference whether the acid or the metal is in excess.
- A change in concentration of the acid will alter the shape of the gas volume graph but the final gas volume will be changed *only if* all the acid is used up (i.e. the metal is in excess).

The gas volume curves for the reaction of hydrochloric acid and zinc are shown in the graph in Figure 9.6 for three different temperatures of acid.

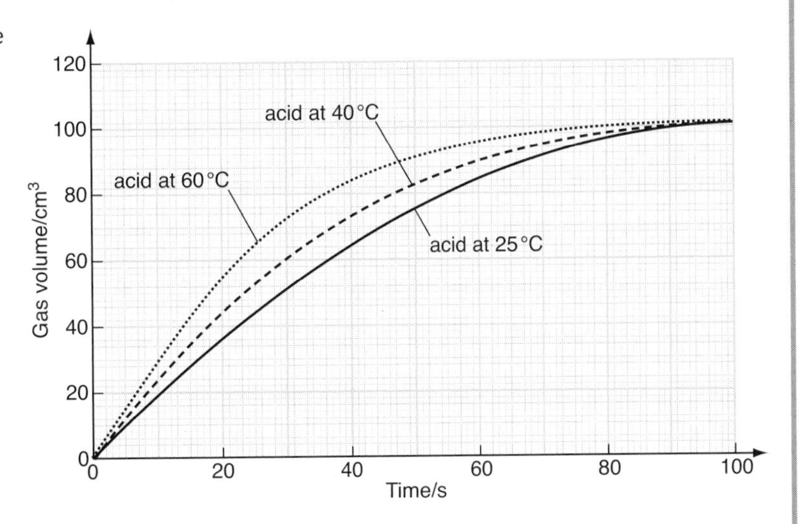

Figure 9.6 Graph of gas volume against time for the reaction of zinc and hydrochloric acid

NOTE: The same four points (page 94) are true for the higher temperature and the converse is true of lower temperatures.

At higher concentrations of acid the same effect is observed. The final gas volume is the same as long as the acid is in excess and all the metal is used up. If the acid is all used up (i.e. metal remains at the end), then the final gas volume changes. You can work this out using a simple mole calculation (page 73).

Typical question

An increase in the concentration of hydrochloric acid increases the rate of reaction of the acid with magnesium ribbon. Explain in terms of particles how this occurs. *[4]*

Answer

As the concentration of acid increases there is an increase in the numbers of hydrogen ions in the acid *[1]* which increases the number of collisions *[1]* which increases the number of successful collisions *[1]* in a given period of time *[1]*

HINT: This question brings up three common errors.

- The name of the reacting particles that increase in concentration: in this question it is an acid – so the reacting particles are hydrogen ions – but if you are unsure just state that 'there is an increase in the number of particles'.
- Both collision and successful collision must be mentioned separately: don't think you can get away with 'there are more successful collisions' for both of the marks.
- The last part of the answer is often left out: you must indicate that an increase in rate is an increase in the number of successful collisions *per second* or in a *given period of time*.

Catalysis

A **catalyst** is a substance that increases the rate of a chemical reaction without being used up.

Table 9.1 lists some common catalysts.

Table 9.1 Common catalysts and their applications

Reaction/industrial process	Catalyst
decomposition of hydrogen peroxide	manganese(IV) oxide, MnO_2
Contact process (H_2SO_4 manufacture)	vanadium(V) oxide, V_2O_5
Haber-Bosch process (NH_3 manufacture)	iron
Nitric acid manufacture	platinum/rhodium

Enzymes are **proteins** that act as **biological catalysts**. They work at an optimum temperature, usually 37 °C and an optimum pH (usually around 7). Some enzymes can work at lower pH values, such as those which digest protein in the acidic conditions of the stomach.

Typical question

What is meant by the term 'catalyst'? *[3]*

Answer

A substance that speeds up *[1]* a (chemical) reaction *[1]* without being used up itself *[1]*

HINT: The main problem with this question is the last part. Students answer the first two parts well but the most common error is to write 'without taking part in the reaction' for the last part of the answer. This is not the case as the catalyst *does* take part in the reaction but is simply not used up during it (unlike the reactants).

Industrial processes

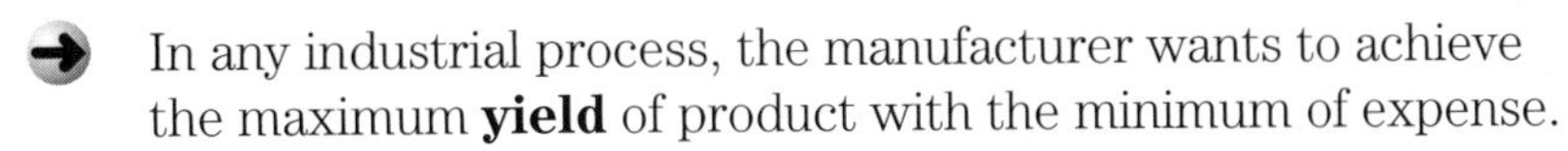

In any industrial process, the manufacturer wants to achieve the maximum **yield** of product with the minimum of expense.

- Often the use of a high temperature, a high pressure and a catalyst can increase the yield of the product.
- However, these factors can be expensive and a **cost analysis** must be carried out to make sure there are significant benefits in terms of profit to justify the expense.
- High pressure is expensive to apply and expensive in terms of the containers needed and it is also dangerous.

The rate of a chemical reaction increases with temperature so if a manufacturer uses a lower temperature, then the rate is lowered even if more of the product is obtained (but over a longer time period).

A **compromise temperature** is used that allows enough product to be made in as short a time period as is possible.

Photographic film

Photographic film contains silver ions, Ag^+, which are converted to silver atoms, Ag, when light falls on the film.

The silver ions gain electrons, $Ag^+ + e^- \rightarrow Ag$. This is a **reduction reaction** (gain in electrons) as the Ag^+ ions gain electrons.

Light increases the rate of conversion of silver ions to silver atoms. Photosynthesis is another reaction that is speeded up by light.

Revision questions

1 Name the catalyst used in the decomposition of hydrogen peroxide. *[1]*

2 In the reaction of sodium thiosulphate with hydrochloric acid, what product causes the solution to become cloudy? *[1]*

3 What piece of apparatus is used to measure gas volume? *[1]*

4 What catalyst is used for the industrial production of ammonia in the Haber-Bosch process? *[1]*

5 State the effect of an increase in temperature on the rate of a chemical reaction. *[1]*

6 Write an equation to represent the reaction of silver(I) ions in photographic film when exposed to light. *[2]*

7 What gas is produced when magnesium reacts with dilute hydrochloric acid? *[1]*

8 Apart from temperature and concentration, name **one** other factor which affects the rate of a chemical reaction. *[1]*

9 Describe briefly what method you would use to determine the rate of reaction for marble chips reacting with dilute hydrochloric acid. *[2]*

10 The production of ammonia uses a temperature of 450 °C and a pressure of 250 atm. A higher yield can be obtained using a lower temperature and a higher pressure.

a Explain why a temperature of 450 °C is used. *[2]*

b Explain why a pressure of 250 atm is used. *[2]*

Questions **11** to **15** relate to the graph shown below.

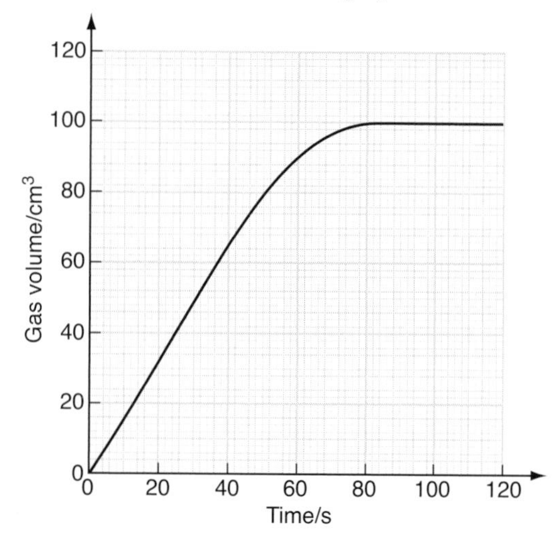

The graph shows the volume of hydrogen gas produced against time for the reaction of zinc with 25.0 cm^3 of 0.5 mol/dm^3 dilute hydrochloric acid at 25 °C. All the zinc was used up in the reaction.

11 At what time did the reaction end? *[1]*

12 What volume of gas had been produced at 50 seconds? *[1]*

13 Copy the graph and then sketch on it the line you would expect to obtain if the temperature was increased to 40 °C, and all other factors were kept the same. Label this 'A'. *[4]*

14 If the concentration of the acid was reduced to 0.4 mol/dm^3 and the same mass of zinc was added at 25 °C, sketch a second line on your graph that you would expect to obtain if all the zinc is again used up. Label this 'B'. *[3]*

15 What volume of gas would be produced if enough zinc was added so that all the 25.0 cm^3 of 0.5 mol/dm^3 hydrochloric acid were used up? *[2]*

10

Metals and their compounds

Physical properties of metals

The following terms are used to describe the physical properties of metals. You need to be able to describe what the terms mean and, for some of them, explain why metals show these properties (see also page 30).

- Metals are **lustrous** (shiny), **malleable** (can be hammered into shape), **ductile** (can be drawn out into wires), **sonorous** (ring when struck), **electrical conductors**, **heat conductors** and have **high melting points**.
- Metals are malleable and ductile as the layers of positive ions can slide over each other without disrupting the bonding.
- Metals conduct electricity as the delocalised electrons can move and carry charge.
- Metals have high melting points as the metallic bonds are strong and require a lot of energy to break them.

Uses of metals related to their properties

Table 10.1 (overleaf) shows some uses of metals.

NOTE: The uses of metals need to be learned and also, where appropriate, how the properties of the metal relate to their use. If the 'property related to use' box is blank, it is beyond what is expected at GCSE level.

Table 10.1 Uses of metals related to their properties

Metal	Uses	Property related to use
calcium	important in the body (healthy bones and teeth)	
magnesium	alloys for aircraft	strong
	flares	burns with a bright white light
aluminium	electrical wiring	conducts electricity; ductile
	saucepans	conducts heat
	alloys	strong
zinc	galvanising	more reactive than iron
	brass	strong
iron	red blood cells	
	structures	strong
copper	electrical wiring	conducts electricity; ductile
	plumbing	malleable
	brass	strong; lack of reactivity
	coinage	lack of reactivity
lead	roofing	malleable; soft
	batteries	
	solder	soft; low melting point
	anti-knock in petrol	

Chemical properties of metals

The **reactivity series** (page 109) of metals predicts the vigour of reactions of metals with:

- air (oxygen)
- water (including steam)
- acids.

Reaction with air

Table 10.2 shows the reactions of metals with air (oxygen).

Table 10.2 Reactions of metals with air (oxygen)

Metal	Reaction when heated in air	Reaction with air under ambient conditions	Equation
K	burns with a lilac flame	when freshly cut, the shiny surface tarnishes (goes dull) rapidly	$4K + O_2 \rightarrow 2K_2O$
Na	burns with a golden yellow flame		$4Na + O_2 \rightarrow 2Na_2O$
Ca	burns with a red flame	react slowly forming a surface oxide layer	$2Ca + O_2 \rightarrow 2CaO$
Mg	burns with a bright white light, producing a white powder		$2Mg + O_2 \rightarrow 2MgO$
Al	burns *only when a fine powder*, forming white solid aluminium oxide		$4Al + 3O_2 \rightarrow 2Al_2O_3$
Zn	burns in air, forming solid zinc oxide which appears yellow but changes to white on cooling		$2Zn + O_2 \rightarrow 2ZnO$
Fe	iron filings burn like a sparkler, forming black mixed oxide, tri-iron tetroxide, Fe_3O_4		$3Fe + 2O_2 \rightarrow Fe_3O_4$
Cu	forms black copper oxide without burning		$2Cu + O_2 \rightarrow 2CuO$

Sodium, potassium and calcium are only heated in air under very careful supervision and strict safety procedures. The reactions can be extremely dangerous.

All other metals listed in Table 10.2 can be heated in air in a crucible using the apparatus shown in Figure 10.1.

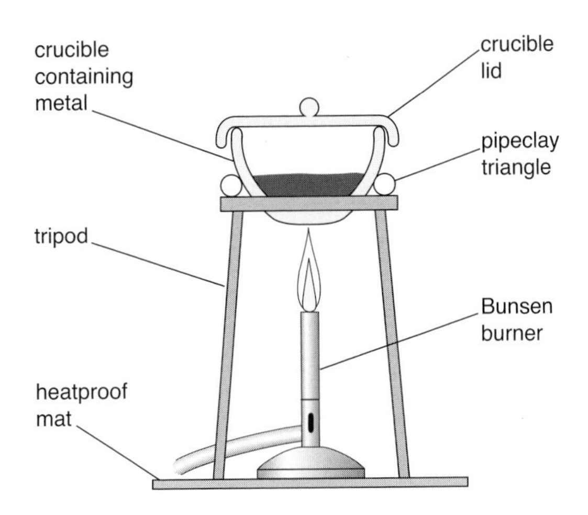

Figure 10.1 Apparatus used to heat metals (except sodium, potassium, calcium) in air

Usually the powder form of the metal is heated. The crucible lid is lifted slightly during heating to allow more air to get into the crucible.

When metals react with air they gain mass as they form an oxide.

Reaction with water

Table 10.3 shows the reactions of metals with water.

Table 10.3 Reactions of metals with water

Metal	Reaction with water	Equation and notes
K	• floats on the surface • moves about the surface • burns with a lilac flame • fizzes, gas given off • eventually disappears • heat given out • colourless solution formed	$2K + 2H_2O \rightarrow 2KOH + H_2$ (potassium is stored under oil to prevent reaction with oxygen and moisture in air)
Na	• floats on surface • melts and forms a silvery ball • moves about the surface • fizzes, gas given off • eventually disappears • heat given out • colourless solution formed	$2Na + 2H_2O \rightarrow 2NaOH + H_2$ (sodium is stored under oil to prevent reaction with oxygen and moisture in air)
Ca	• fizzes, gas given off • eventually disappears • heat given out • colourless solution formed	$Ca + 2H_2O \rightarrow Ca(OH)_2 + H_2$
Mg	• very slow reaction, producing a few bubbles of gas	$Mg + 2H_2O \rightarrow Mg(OH)_2 + H_2$

When sodium and potassium are reacted with water, a small piece is placed in a trough half-full of water. This is carried out behind a safety screen.

To react calcium or magnesium with water, the metal is put in a beaker with water and an inverted filter funnel is placed on top with a boiling tube filled with water to collect the hydrogen produced (Figure 10.2).

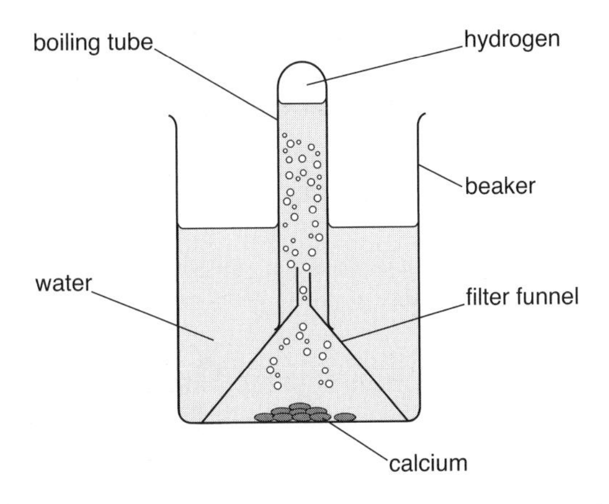

Figure 10.2 Apparatus used to react metals (except sodium, potassium) with water

When magnesium is used, only a few bubbles are produced over the period of a few days.

Reaction with steam

Table 10.4 shows the reactions of metals with steam.

Table 10.4 Reactions of metals with steam

Metal	Reaction with steam	Equation
Mg	reacts when heated, producing magnesium oxide and hydrogen gas: • bright white light • white powder remains • heat given out	$Mg + H_2O \rightarrow MgO + H_2$
Al	no reaction in foil form – aluminium only reacts when the protective layer of aluminium oxide is removed; powdered form reacts, producing white aluminium oxide	$2Al + 3H_2O \rightarrow Al_2O_3 + 3H_2$
Zn	reacts when heated, forming zinc oxide and hydrogen gas: • glows • yellow powder which changes to white on cooling • heat given out	$Zn + H_2O \rightarrow ZnO + H_2$
Fe	reacts reversibly at red heat, forming tri-iron tetroxide: • black solid forms	$3Fe + 4H_2O \rightarrow Fe_3O_4 + 4H_2$
Cu	no reaction	no reaction

The apparatus shown in Figure 10.3 is used to allow steam to react with a heated metal. The apparatus is connected to a delivery tube and the gas produced collected over water using a beehive shelf and a gas jar. The gas produced is hydrogen.

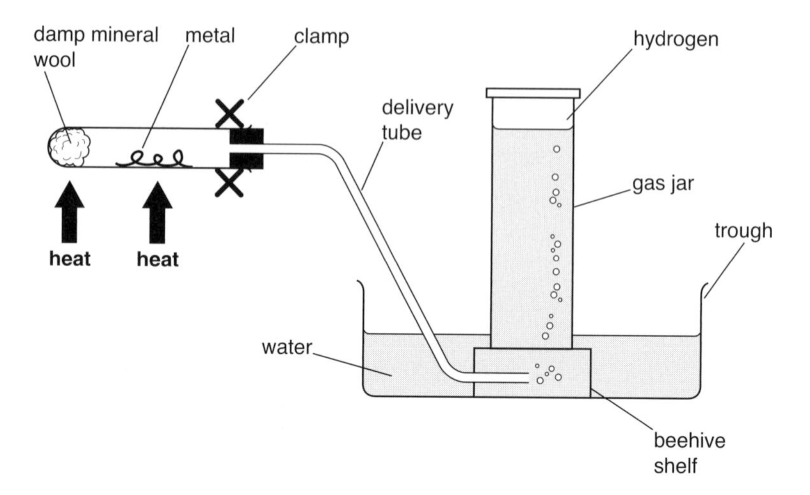

Figure 10.3 Apparatus used to react metals with steam

The damp mineral wool is heated to generate steam. When the heating stops, there is a risk of suck back occurring (suck back is the water in the trough being drawn back into the hot boiling tube). This can be prevented by removing the apparatus from the water in the trough or by taking out the bung in the boiling tube.

HINT: When drawing a diagram in answer to a question, marks are awarded for the labels on an assembled and recognisable diagram. Make the apparatus look like it should. For example, you will *not* gain a mark for a beaker labelled as a crucible.
The most common mistake when drawing the diagram in Figure 10.3 is to put the delivery tube through the wall of the trough. Remember the level of the water in the trough should always be *above* the beehive shelf.

Reaction with acids

Table 10.5 shows the reactions of metals with dilute acids.

Table 10.5 Reactions of metals with dilute acids

Metal	Reaction with dilute hydrochloric/sulphuric acid	Equations
K Na	dangerous, violent reaction (not attempted in the laboratory)	
Ca	• bubbles of gas produced • heat given out • colourless solution formed • metal eventually disappears	$Ca + 2HCl \rightarrow CaCl_2 + H_2$ $Ca + H_2SO_4 \rightarrow CaSO_4 + H_2$
Mg	• bubbles of gas produced • heat given out • colourless solution formed • metal eventually disappears	$Mg + 2HCl \rightarrow MgCl_2 + H_2$ $Mg + H_2SO_4 \rightarrow MgSO_4 + H_2$
Al	foil reacts after a few minutes due to protective oxide layer; powdered aluminium reacts immediately: • bubbles of gas produced • heat given out • colourless solution formed	$2Al + 6HCl \rightarrow 2AlCl_3 + 3H_2$ $2Al + 3H_2SO_4 \rightarrow Al_2(SO_4)_3 + 3H_2$
Zn	• bubbles of gas produced • heat given out • colourless solution formed • metal eventually disappears	$Zn + 2HCl \rightarrow ZnCl_2 + H_2$ $Zn + H_2SO_4 \rightarrow ZnSO_4 + H_2$
Fe	• bubbles of gas produced • heat given out • very pale green solution formed • metal eventually disappears	$Fe + 2HCl \rightarrow FeCl_2 + H_2$ $Fe + H_2SO_4 \rightarrow FeSO_4 + H_2$
Cu	no reaction	no reaction

NOTE: The disappearance of the metal will depend on the mass of metal used and the volume and concentration of the acid but, for normal laboratory experiments, the metal usually disappears.

Figure 10.4 shows the apparatus used to react a metal with a dilute acid and to collect the hydrogen gas produced.

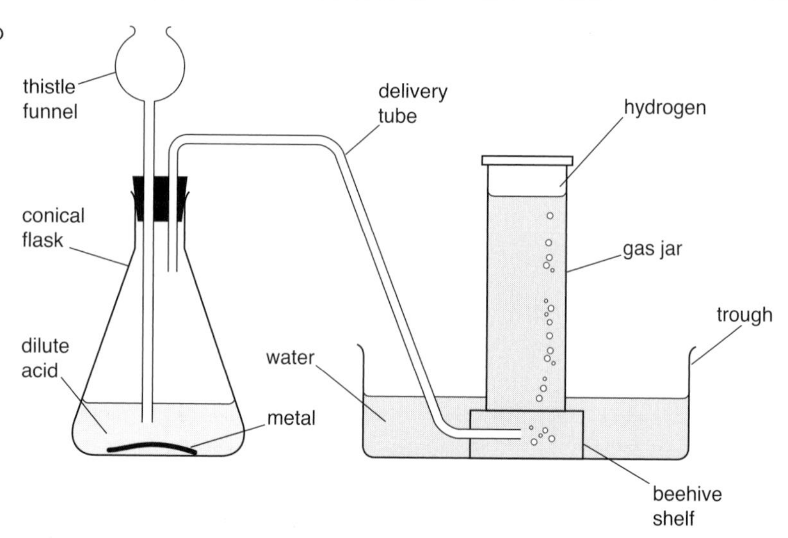

Figure 10.4 Apparatus used to react metals with a dilute acid

Iron produces iron(II) compounds with dilute hydrochloric acid and dilute sulphuric acid. Note that iron reacts with chlorine gas to produce iron(III) chloride, but that with hydrogen chloride gas or hydrochloric acid, iron(II) chloride is produced (page 105).

Aluminium reacts very slowly with dilute hydrochloric acid and virtually not at all with dilute sulphuric acid but will react faster if the acid is more concentrated and hot.

Typical question

What is observed when magnesium reacts with dilute hydrochloric acid? *[4]*

Answer

Bubbles of gas produced *[1]*; heat *[1]*; magnesium disappears *[1]*; colourless solution *[1]*

HINT: There are two common mistakes made in answering this question. Simply stating that 'gas is produced' may not be enough to gain the first mark as the actual observation must include the word 'bubbles'. The second common error is stating that magnesium dissolves. Dissolving is a physical change and this is a chemical reaction so the answer must be that 'magnesium *disappears*'.

Displacement reactions

A **displacement reaction** is one where a **more reactive metal** forms an ion and causes a **less reactive metal ion** to change to atoms. The process involves the transfer of electrons.

- There are two main types of displacement reaction:
  - solid metal reacting with a solution of a metal ion
  - solid metal reacting with a solid metal oxide.
- One species loses electrons and one gains electrons.
  - loss of electrons is called **oxidation**
  - gain of electrons is called **reduction**.
- When both oxidation and reduction reactions occur in the same reaction, it is described as a **redox reaction**.
- Oxidation and reduction can also be defined in terms of loss and gain of oxygen or gain or loss of hydrogen. Table 10.6 gives all the definitions of oxidation and reduction.

Table 10.6 The definitions of oxidation and reduction

Oxidation	Reduction
gain of oxygen	loss of oxygen
loss of hydrogen	gain of hydrogen
loss of electrons	gain of electrons

Solutions

Example 1

Magnesium metal reacts when placed in a solution of copper(II) sulphate. What type of reaction is this?

Observations: blue solution fades in colour to colourless; red/pink or black solid appears; heat given out

Symbol equation: $Mg + CuSO_4 \rightarrow MgSO_4 + Cu$

Ionic equation: $Mg + Cu^{2+} \rightarrow Mg^{2+} + Cu$

Spectator ion: SO_4^{2-} (does not take part in the reaction and is the same in reactants and products)

Half equations: $Mg \rightarrow Mg^{2+} + 2e^-$ *oxidation*

magnesium atoms lose electrons and loss of electrons is oxidation

$Cu^{2+} + 2e^- \rightarrow Cu$ *reduction*

copper(II) ions gain electrons and gain of electrons is reduction

This reaction is a **redox reaction** as both oxidation and reduction are occurring simultaneously (see page 162).

Extraction of metals from their ores

- The lowest reactivity metals are found uncombined in nature (also called **native**).
- Metals that are high up in the reactivity series are extracted by **electrolysis**.
- Metals that are low in the reactivity series are extracted by **reduction** with carbon.
- For metals such as aluminium and those above it in the reactivity series, electrolysis must be used to extract the metal.

HINT: When describing the reactivity of metals always write for example 'magnesium is more reactive than copper'. A common mistake is to state 'magnesium is more reactive', but you must say more reactive *than* the other metal.

In the laboratory, powdered carbon can be mixed with the metal oxide and heated in a crucible. The reduction of the metal oxide occurs side-by-side with the oxidation of the carbon to carbon monoxide or sometimes carbon dioxide. For example:

$$ZnO + C \rightarrow Zn + CO$$

This is a **redox** reaction because:

- the zinc ions gain electrons to form Zn atoms and this is reduction
- the carbon gains oxygen and this is oxidation
- oxidation and reduction are occurring simultaneously.

Reaction of metal oxides with hydrogen gas

As we have seen in the previous section, hydrogen gas can be used to reduce metal oxides to the metal for metals like copper and ones below it in the reactivity series (page 109). Usually copper oxide is used (Figure 10.5).

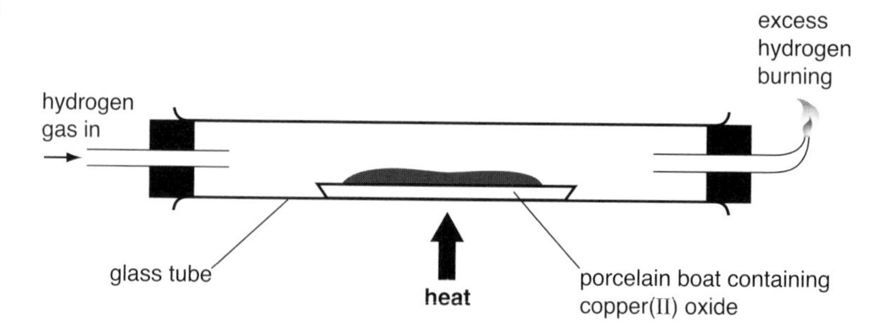

Figure 10.5 Apparatus used to react a metal oxide with hydrogen gas

Observations: black copper oxide changes to red/pink; condensation appears on inside of glass tube

Equation: $CuO + H_2 \rightarrow CuO + H_2O$

This is a **redox** reaction because:

- copper ions in the copper(II) oxide gain electrons and gain of electrons is reduction
- hydrogen gains oxygen and gain of oxygen is oxidation
- oxidation and reduction are occurring simultaneously.

HINT: The description of a reaction as redox is used throughout this section as it often occurs in a question. You should be able to explain in terms of electrons, oxygen and hydrogen why a reaction involves both oxidation and reduction.

Other reactions of metals and their compounds

The reactions of the metals sodium, zinc, aluminium, calcium, copper and iron are described below.

Sodium hydroxide, NaOH

1 Reaction of sodium hydroxide solution with carbon dioxide, CO_2.

Observation: white crust forms (= sodium carbonate)

Equation: $2NaOH + CO_2 \rightarrow Na_2CO_3 + H_2O$

2 Reaction of sodium hydroxide with ammonium compounds (chloride or sulphate).

Observations: on heating produces a colourless pungent gas (ammonia)

Equations: $NH_4Cl + NaOH \rightarrow NaCl + NH_3 + H_2O$
$(NH_4)_2SO_4 + 2NaOH \rightarrow Na_2SO_4 + 2NH_3 + 2H_2O$

This reaction is best carried out by mixing solid sodium hydroxide with the solid ammonium salt and heating them in a boiling tube.

3 Reaction of sodium hydroxide solution with metal ions.

- Aluminium ion, Al^{3+}

  Observations: white precipitate (ppt) which dissolves as more NaOH is added, forming a colourless solution

  Equation: $Al^{3+} + 3OH^- \rightarrow Al(OH)_3$
  white ppt

- Calcium ion, Ca^{2+}

  Observations: white precipitate which does not dissolve as more NaOH is added

  Equation: $Ca^{2+} + 2OH^- \rightarrow Ca(OH)_2$
  white ppt

- Copper(II) ion, Cu^{2+}

  Observations: pale blue precipitate which does not dissolve as more NaOH is added

  Equation: $Cu^{2+} + 2OH^- \rightarrow Cu(OH)_2$
  blue ppt

- Iron(II) ion, Fe^{2+}

  Observations: pale green precipitate which does not dissolve as more NaOH is added

  Equation: $Fe^{2+} + 2OH^- \rightarrow Fe(OH)_2$
  pale green ppt

- Iron(III) ion, Fe^{3+}

  Observations: reddish-brown precipitate which does not dissolve as more NaOH is added

  Equation: $Fe^{3+} + 3OH^- \rightarrow Fe(OH)_3$
  red/brown ppt

- Zinc ion, Zn^{2+}

  Observations: white precipitate which dissolves as more NaOH is added, forming a colourless solution

  Equation: $Zn^{2+} + 2OH^- \rightarrow Zn(OH)_2$
  white ppt

4 Reaction of sodium hydroxide solution with aluminium hydroxide and zinc hydroxide (see page 46).

- The oxides and hydroxides of some metals (i.e. Al and Zn) react with both acids (as expected) and alkalis to give a salt and water only.
- Metal hydroxides that react with both acids and alkalis are called **amphoteric**.
- Zinc hydroxide can produce salts called **zincates**; aluminium hydroxide can produce salts called **aluminates**.

Zinc hydroxide

1 Reaction of zinc hydroxide with acid

$Zn(OH)_2$	+	H_2SO_4	→	$ZnSO_4$	+	$2H_2O$
zinc hydroxide		sulphuric acid		zinc sulphate		water

2 Reaction of zinc hydroxide with alkali

$Zn(OH)_2$	+	$2NaOH$	→	Na_2ZnO_2	+	$2H_2O$
zinc hydroxide		sodium hydroxide		sodium **zincate**		water

Aluminium hydroxide

1 Reaction of aluminium hydroxide with acid

$2Al(OH)_3$	+	$3H_2SO_4$	→	$Al_2(SO_4)_3$	+	$6H_2O$
aluminium hydroxide		sulphuric acid		aluminium sulphate		water

2 Reaction of aluminium hydroxide with alkali

$Al(OH)_3$	+	$NaOH$	→	$NaAlO_2$	+	$2H_2O$
aluminium hydroxide		sodium hydroxide		sodium **aluminate**		water

HINT: The zincate ion is ZnO_2^{2-} and the aluminate ion is AlO_2^-. (These can be written in different ways but this is sufficient for GCSE.) The equations here must be learned thoroughly.

Aluminium hydroxide is used as an **antacid**. An antacid is used to neutralise excess acid in your stomach when you have indigestion.

Aluminium

Aluminium metal also reacts with both acids and alkalis and these equations are required as well.

1 Reaction of aluminium with acid

$2Al$	+	$6HCl$	→	$2AlCl_3$	+	$3H_2$
aluminium		hydrochloric acid		aluminium chloride		hydrogen

2 Reaction of aluminium with alkali

$2Al$	+	$2NaOH$	+	$2H_2O$	→	$2NaAlO_2$	+	$3H_2$
		sodium hydroxide				sodium **aluminate**		hydrogen

Sodium carbonate (hydrated and anhydrous)

- Hydrated sodium carbonate is usually seen as large crystals of a translucent white solid.
- Its common name is washing soda (page 61).
- Its formula is $Na_2CO_3.10H_2O$.
- It is soluble in water.

1 On heating in a boiling tube, it loses its water of crystallisation and forms a white powder which is anhydrous sodium carbonate, Na_2CO_3. Condensation is seen at the top of the boiling tube. There is a loss in mass.

$$Na_2CO_3.10H_2O \rightarrow Na_2CO_3 + 10H_2O$$

2 Solid hydrated sodium carbonate and anhydrous sodium carbonate react with dilute acids in the same way.

Observations: bubbles of gas produced; heat released; solid disappears

Equations: $Na_2CO_3 + 2HCl \rightarrow 2NaCl + CO_2 + H_2O$
$Na_2CO_3 + H_2SO_4 \rightarrow Na_2SO_4 + CO_2 + H_2O$

Sodium hydrogen carbonate

- Sodium hydrogen carbonate is a white solid that has the formula $NaHCO_3$.
- It is soluble in water.
- It is used as an **antacid** (which neutralises excess stomach acid helping indigestion) and in cooking (produces carbon dioxide to make some breads rise).

1 Sodium hydrogen carbonate decomposes on heating to form sodium carbonate, carbon dioxide and water.

Observations: no apparent observations, but there is a loss in mass

Equation: $2NaHCO_3 \rightarrow Na_2CO_3 + H_2O + CO_2$

2 Solid sodium hydrogen carbonate reacts with dilute hydrochloric acid and sulphuric acid.

Observations: bubbles of gas produced; heat released; solid disappears

Equations: $NaHCO_3 + HCl \rightarrow NaCl + CO_2 + H_2O$
$2NaHCO_3 + H_2SO_4 \rightarrow Na_2SO_4 + 2CO_2 + 2H_2O$

Sodium chloride

- Sodium chloride (NaCl) is a white solid and is also known as common salt.
- It is the salt which is used to flavour food and this is one of its main uses.
- It is also used as a preservative for some foods and to de-ice roads in winter.
- It is extracted from rocks below the ground by pumping down very hot water in which the sodium chloride dissolves. The solution returns to the surface and the water is evaporated to leave sodium chloride. This whole process is called solution mining.

Calcium carbonate

- Calcium carbonate is a white solid with the formula $CaCO_3$.
- It is the main chemical in limestone and marble.
- It is insoluble in water.
- It is used in cement manufacture, for removing impurities from the iron ore in the blast furnace and in blackboard chalk.

1 On heating, calcium carbonate decomposes to form calcium oxide and carbon dioxide.

Observations: orange glow and a loss in mass

Equation: $CaCO_3 \rightarrow CaO + CO_2$

Limestone (calcium carbonate) is decomposed into lime (calcium oxide) industrially in a **limekiln** (Figure 10.6). Hot air is blown into the limekiln to burn the coke and heat the calcium carbonate until it decomposes. The exhaust gases mainly consist of carbon dioxide.

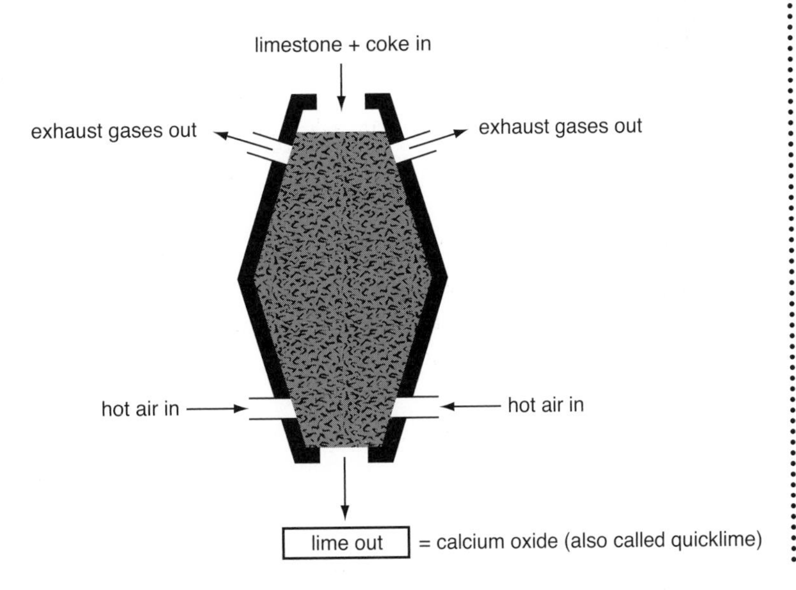

Figure 10.6 A limekiln

2 Calcium carbonate reacts with hydrochloric acid.

Observations: bubbles of gas produced; heat released; solid disappears

Equation: $CaCO_3 + 2HCl \rightarrow CaCl_2 + CO_2 + H_2O$

3 Calcium carbonate reacts initially with dilute sulphuric acid but the surface becomes coated with insoluble calcium sulphate and the reaction stops after a few seconds.

Observations: bubbles of gas produced; reaction stops after a few seconds

Equation: $CaCO_3 + H_2SO_4 \rightarrow CaSO_4 + CO_2 + H_2O$

Calcium sulphate, $CaSO_4$, is used in Plaster of Paris, blackboard chalk and for white line markings on roads.

Calcium oxide

- Calcium oxide is a white solid.
- For the reactions of calcium oxide, a fresh sample should be prepared by heating calcium carbonate (as calcium oxide reacts with the water vapour in the atmosphere).

1 Calcium oxide reacts with water to form a solution of calcium hydroxide.

Observations: hissing; expands; crumbles; heat is released

Equation: $CaO + H_2O \rightarrow Ca(OH)_2$

2 Calcium oxide reacts with acids, forming a salt and water.

Observations: heat released; hissing; colourless solution is formed

Equations: $CaO + 2HCl \rightarrow CaCl_2 + H_2O$
$CaO + H_2SO_4 \rightarrow CaSO_4 + H_2O$

Calcium hydroxide solution (limewater)

- Calcium hydroxide is a white solid that is partially soluble in water forming a colourless solution.
- A solution of calcium hydroxide is called **limewater**.
- Solid calcium hydroxide is used in agriculture to neutralise acidic soils.
- Limewater is used to test for carbon dioxide.
- When carbon dioxide reacts with limewater, a white precipitate of $CaCO_3$ forms which causes 'milkiness' in the solution.

Carbon dioxide reacts with limewater.

Observations: colourless solution changes to milky (caused by a white precipitate of calcium carbonate)

Equation: $Ca(OH)_2 + CO_2 \rightarrow CaCO_3 + H_2O$
limewater white ppt

When carbon dioxide is in excess, the limewater changes back to a colourless solution.

Equation: $CaCO_3 + CO_2 + H_2O \rightarrow Ca(HCO_3)_2$
white ppt colourless solution

Calcium hydrogen carbonate solution

- Calcium hydrogen carbonate solution is a colourless solution.
- Calcium hydrogen carbonate in solution is the cause of temporary hardness in water (page 60).

Heating the solution causes the calcium hydrogen carbonate to decompose to form calcium carbonate.

Observations: a white precipitate forms in a colourless solution

Equation: $Ca(HCO_3)_2 \rightarrow CaCO_3 + CO_2 + H_2O$
white ppt

Copper(II) carbonate

- Copper(II) carbonate (normally called copper carbonate) is a green solid.

1 When copper carbonate is heated it decomposes to form copper oxide and carbon dioxide gas is produced.

Observations: green solid changes to black

Equation: $CuCO_3 \rightarrow CuO + CO_2$

2 When copper carbonate is added to a dilute acid (hydrochloric or sulphuric) a blue solution forms and carbon dioxide gas is produced.

Observations: bubbles of gas produced; heat released; green solid disappears; blue solution formed

Equations: $CuCO_3 + 2HCl \rightarrow CuCl_2 + CO_2 + H_2O$
$CuCO_3 + H_2SO_4 \rightarrow CuSO_4 + CO_2 + H_2O$

Copper(II) sulphate

- Copper(II) sulphate can be hydrated or anhydrous.
- When hydrated it is blue and dissolves in water to form a blue solution.
- When anhydrous it is white but still dissolves in water to form a blue solution.

1 Heating hydrated copper(II) sulphate in a boiling tube removes the water of crystallisation.

Observations: changes colour from blue to white and condensation appears on the inside of the boiling tube

Equation: $CuSO_4.5H_2O \rightarrow CuSO_4 + 5H_2O$

2 When water is added to anhydrous copper(II) sulphate, it changes from white to blue. This is used as a test for water.

Iron(II) and iron(III) compounds

- Iron compounds can either contain the iron(II) ion, Fe^{2+}, or the iron(III) ion, Fe^{3+}.
- Fe^{2+} can be **oxidised** to Fe^{3+} using an **oxidising agent**. $Fe^{2+} \rightarrow Fe^{3+} + e^-$. The Fe^{2+} loses 1 electron to the oxidising agent and loss of electrons is oxidation.
- Air and chlorine oxidise Fe^{2+} to Fe^{3+}.
- Iron(II) compounds are pale green and if they are soluble, they form pale green solutions.
- Iron(III) compounds are usually red-brown and if they are soluble, they form yellow solutions.

1 Iron(II) reacts with chlorine (either chlorine water or chlorine gas).

Observations: green solution changes to yellow solution

Equation: $2Fe^{2+} + Cl_2 \rightarrow 2Fe^{3+} + 2Cl^-$

2 Iron(II) ions reacts with oxygen in air to form iron(III) ions.

Observations: green solution changes to yellow solution

Equation: $4Fe^{2+} + O_2 \rightarrow 4Fe^{3+} + 2O^{2-}$

Rusting

Rust is **hydrated iron(III) oxide**, sometimes written $Fe_2O_3.xH_2O$. When iron is exposed to air and moisture (water in the air), it rusts (see page 168).

An investigation to determine the factors that cause rusting is shown in Figure 10.7.

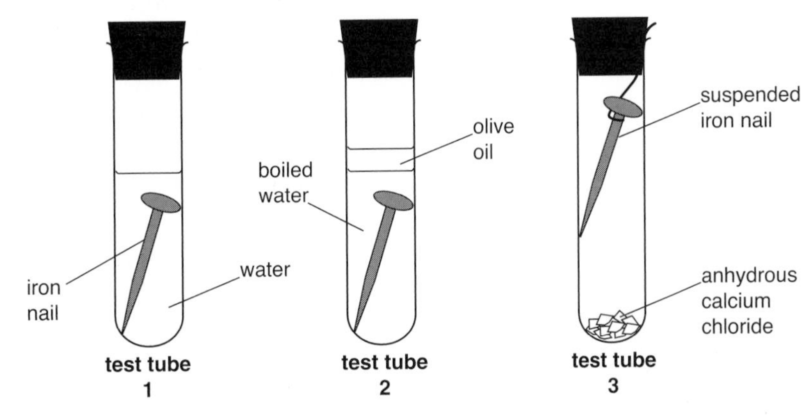

Figure 10.7 Apparatus to investigate the factors that cause rusting

- Test tube **1** has air and water present.
- Test tube **2** has had the air removed from the water so only water is present. The olive oil prevents gases in the air dissolving in the water.
- Test tube **3** contains anhydrous calcium chloride that removes the water from the air so only air is present. The nail is suspended to stop contact between it and the calcium chloride.

The test tubes are left for several days and the iron nail rusts *only* in test tube **1**. This indicates that air and water are both required for rusting.

Prevention of rusting

Iron is used extensively in construction and rusting is a major problem due to the cost of replacing the iron. Rust is unsightly and dangerous as it weakens the iron. Rusting can be prevented in a variety of ways.

Methods to prevent rust formation fall into three groups.

1 Preventing the surface of the iron coming into contact with water and air by using a barrier or protective layer:
   - paint is used to protect cars, bridges and railings
   - oil or grease is used to protect tools and machinery
   - plastic coating such as that used to cover bicycle handlebars, garden chairs and dish racks. Car manufacturers are increasingly using more plastic in cars to reduce the problem of rust

- plating with another metal:
  - tin is used in cans of food. These cans are made from steel and are coated on both sides with a thin layer of tin. Tin is unreactive and non-toxic. It is deposited on the steel by electrolysis
  - chromium can be used to coat steel giving an attractive, shiny appearance. This is sometimes used for vehicle bumpers and bicycle handlebars. Chromium can be plated by electrolysis.

2 Placing a more reactive metal in contact with the iron or steel. The more reactive metal reacts first, leaving the iron intact.

- Bars of magnesium are attached to the sides of ships, oil rigs and underwater pipes to prevent rust. The magnesium **corrodes** instead of the steel and must be periodically replaced with fresh magnesium. This method of rust prevention is called **sacrificial protection**.
- Steel is coated in zinc. This is called **galvanising**. Zinc is more reactive than iron and readily oxidises to form a layer of zinc oxide. Galvanising protects by sacrificial protection if the surface is scratched, and also the zinc oxide provides a barrier to air and water.

3 Combining two or more metals, called **alloying**. An **alloy** is a mixture of two or more metals. Alloys are often stronger and more resistant to corrosion than the pure metals. Stainless steel is an alloy that is resistant to corrosion.

Extraction of iron in the blast furnace

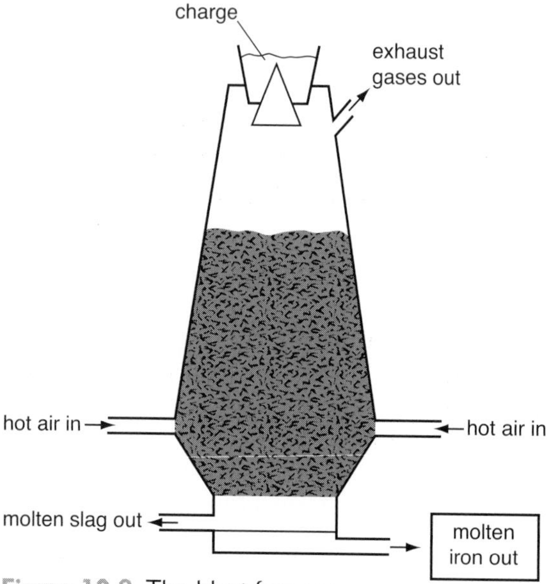

Figure 10.8 The blast furnace

Figure 10.8 shows a simplified diagram of the blast furnace. The solid material added to the blast furnace is called the **charge**. It is made up of **iron ore** (haematite, Fe_2O_3), **limestone** (calcium carbonate, $CaCO_3$) and **coke** (carbon, C). Hot air is blasted in through pipes near the bottom of the blast furnace. Reduction of iron ore happens as the oxygen is removed from the iron(III) oxide *or* Fe^{3+} ions gain electrons (reduction) to form Fe.

Rusting

Rust is **hydrated iron(III) oxide**, sometimes written $Fe_2O_3.xH_2O$. When iron is exposed to air and moisture (water in the air), it rusts (see page 168).

An investigation to determine the factors that cause rusting is shown in Figure 10.7.

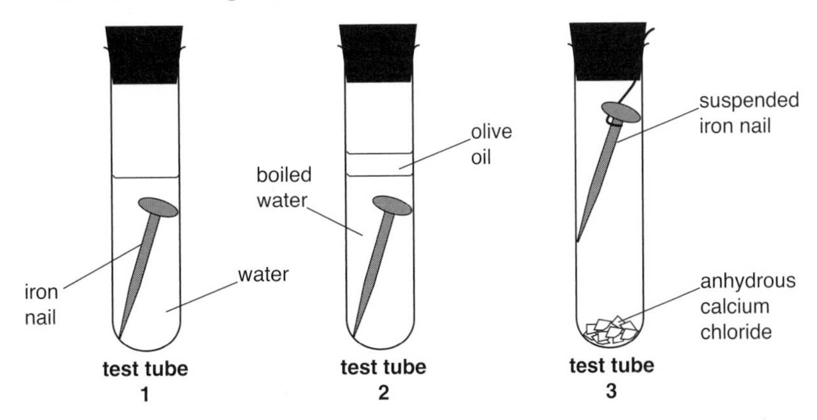

Figure 10.7 Apparatus to investigate the factors that cause rusting

- Test tube **1** has air and water present.
- Test tube **2** has had the air removed from the water so only water is present. The olive oil prevents gases in the air dissolving in the water.
- Test tube **3** contains anhydrous calcium chloride that removes the water from the air so only air is present. The nail is suspended to stop contact between it and the calcium chloride.

The test tubes are left for several days and the iron nail rusts *only* in test tube **1**. This indicates that air and water are both required for rusting.

Prevention of rusting

Iron is used extensively in construction and rusting is a major problem due to the cost of replacing the iron. Rust is unsightly and dangerous as it weakens the iron. Rusting can be prevented in a variety of ways.

Methods to prevent rust formation fall into three groups.

1 Preventing the surface of the iron coming into contact with water and air by using a barrier or protective layer:
   - paint is used to protect cars, bridges and railings
   - oil or grease is used to protect tools and machinery
   - plastic coating such as that used to cover bicycle handlebars, garden chairs and dish racks. Car manufacturers are increasingly using more plastic in cars to reduce the problem of rust

- plating with another metal:
  - tin is used in cans of food. These cans are made from steel and are coated on both sides with a thin layer of tin. Tin is unreactive and non-toxic. It is deposited on the steel by electrolysis
  - chromium can be used to coat steel giving an attractive, shiny appearance. This is sometimes used for vehicle bumpers and bicycle handlebars. Chromium can be plated by electrolysis.

2 Placing a more reactive metal in contact with the iron or steel. The more reactive metal reacts first, leaving the iron intact.

- Bars of magnesium are attached to the sides of ships, oil rigs and underwater pipes to prevent rust. The magnesium **corrodes** instead of the steel and must be periodically replaced with fresh magnesium. This method of rust prevention is called **sacrificial protection**.
- Steel is coated in zinc. This is called **galvanising**. Zinc is more reactive than iron and readily oxidises to form a layer of zinc oxide. Galvanising protects by sacrificial protection if the surface is scratched, and also the zinc oxide provides a barrier to air and water.

3 Combining two or more metals, called **alloying**. An **alloy** is a mixture of two or more metals. Alloys are often stronger and more resistant to corrosion than the pure metals. Stainless steel is an alloy that is resistant to corrosion.

Extraction of iron in the blast furnace

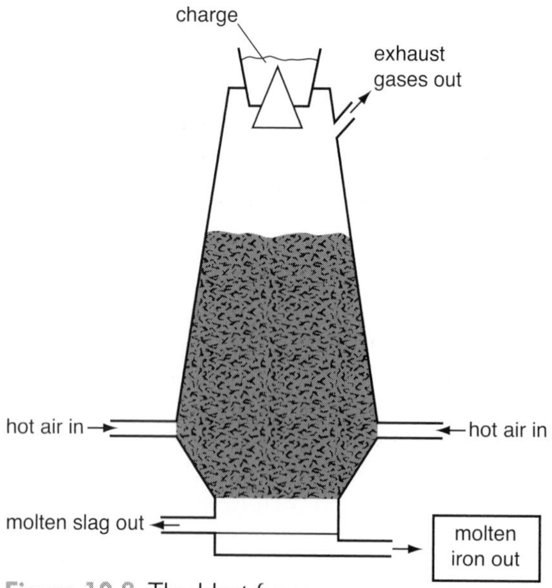

Figure 10.8 shows a simplified diagram of the blast furnace. The solid material added to the blast furnace is called the **charge**. It is made up of **iron ore** (haematite, Fe_2O_3), **limestone** (calcium carbonate, $CaCO_3$) and **coke** (carbon, C). Hot air is blasted in through pipes near the bottom of the blast furnace. Reduction of iron ore happens as the oxygen is removed from the iron(III) oxide *or* Fe^{3+} ions gain electrons (reduction) to form Fe.

Figure 10.8 The blast furnace

Reduction of iron ore in the blast furnace

Stage 1: coke burns in oxygen (from hot air) to produce carbon dioxide

$C + O_2 \rightarrow CO_2$

Stage 2: carbon dioxide reacts with more coke to produce the reducing agent which is carbon monoxide

$CO_2 + C \rightarrow 2CO$

Stage 3: iron(III) oxide reacts with carbon monoxide to produce molten iron and carbon dioxide

$Fe_2O_3 + 3CO \rightarrow 2Fe + 3CO_2$

Removal of impurities from the molten iron

Stage 1: calcium carbonate thermally decomposes to form calcium oxide and carbon dioxide

$CaCO_3 \rightarrow CaO + CO_2$

Stage 2: silicon dioxide impurities react with calcium oxide to remove impurities as molten **slag** (calcium silicate, $CaSiO_3$)

$SiO_2 + CaO \rightarrow CaSiO_3$

The molten slag and molten iron fall to the bottom. Iron is denser than slag so the iron lies below the slag. They are tapped off separately at the base of the blast furnace.

HINT: A lot of the questions about iron involve the materials added to the blast furnace, the materials removed from it and the equations for the five reactions that occur during the process. The most common mistakes are to forget to write *molten* slag and *molten* iron and to get the order in which they are removed mixed up – remember that molten iron is denser than molten slag.

Cation tests

A **cation** is a positive ion. Positive ions occur in ionic compounds and can be tested for in three different ways. You need to know how to perform each test and the expected result for each ion.

Flame test

Method

- dip a flame test rod in deionised water
- place the flame test rod in a Bunsen burner flame to clean it
- place the rod in the sample
- heat sample in Bunsen flame

Results

Table 10.7 shows the results of flame tests for different ions.

Table 10.7 Flame-test colours for different ions

Colour	Ion present
red	Ca^{2+}
blue/green	Cu^{2+}
lilac	K^{+}
golden yellow	Na^{+}

Using sodium hydroxide solution

See page 112 for equations.

Method

Add sodium hydroxide solution to a solution containing the cation.

Results

The colour of the precipitate and whether or not it dissolves in excess sodium hydroxide solution identify the cation present (Table 10.8).

Table 10.8 Determining the cation present using sodium hydroxide solution

Colour of precipitate	Solubility in excess sodium hydroxide	Cation present
white	insoluble	Ca^{2+}
blue	insoluble	Cu^{2+}
white	soluble, forming a colourless solution	Al^{3+}
green	insoluble	Fe^{2+}
red/brown	insoluble	Fe^{3+}
white	soluble, forming a colourless solution	Zn^{2+}

Using aqueous ammonia

See page 133 for equations.

Method

Add aqueous ammonia to a solution containing the metal ion.

Results

The colour of the precipitate and whether or not it dissolves in excess aqueous ammonia identify the cation present (Table 10.9).

Table 10.9 Determining the cation present using aqueous ammonia

Colour of precipitate	Solubility in excess aqueous ammonia	Cation present
blue	soluble, forming a dark blue solution	Cu^{2+}
white	insoluble	Al^{3+}
green	insoluble	Fe^{2+}
red/brown	insoluble	Fe^{3+}
white	soluble, forming a colourless solution	Zn^{2+}
white	insoluble	Mg^{2+}

NOTE: The aqueous ammonia test will distinguish between Al^{3+} and Zn^{2+}.

Determining the reactivity of a metal

➜ **Displacement reactions** can be used to determine a reactivity series.

- A set of reactions between metals and their metal salt solutions is carried out.
- The metals are simply placed in a solution of the metal salt (usually the sulphate). The results are often recorded in a table (Table 10.10, overleaf).
- A tick (✓) is used to indicate a reaction occurring and a cross (✗) indicates no reaction.
- The parts of the table that are shaded show that the metal should not be placed in a solution of its own salt, for example magnesium is not placed in magnesium sulphate solution.
- From the table it is seen that magnesium displaces the other three metals from their solutions, indicating that it is the most reactive.
- Zinc displaces copper and iron from their solutions but does not displace magnesium, which shows it is the next reactive.
- Iron displaces only copper from its solution so iron is the third most reactive.
- Copper does not displace any of the other metals from their solutions so it is the least reactive in this investigation.

The reactivity in order of most reactive to least reactive is:

magnesium zinc iron copper

Table 10.10 Displacement reactions can be used to determine a reactivity series

Metal \ Metal salt solution	magnesium sulphate	copper(II) sulphate	iron(II) sulphate	zinc sulphate
magnesium		✓	✓	✓
copper	✗		✗	✗
iron	✗	✓		✗
zinc	✗	✓	✓	

HINT: Questions often ask for the reactivity in order from most reactive to least reactive but be careful – some ask for the order from least reactive to most reactive.

Equations for the above reactions can be asked for and observations can also be asked. Remember to give the appearance of each substance *before* and *after*. Heat is almost always released in displacement reactions.

Revision questions

1 What does ductile mean? *[1]*

2 What is observed when potassium reacts with water? *[4]*

3 Write a balanced symbol equation for the reaction of potassium with water. *[1]*

4 What is observed when magnesium reacts with copper(II) sulphate solution? *[3]*

5 Write the formulae of **two** metal ions in solution that produce a white precipitate when sodium hydroxide is added to the solution. *[2]*

6 Name the **three** solid substances that are added to the blast furnace in the extraction of iron. *[3]*

7 Write an ionic equation for the reaction of iron(II) ions with hydroxide ions. *[3]*

8 What colours are copper carbonate and copper oxide? *[2]*

9 Write a balanced symbol equation for the thermal decomposition of calcium carbonate. *[2]*

10 A sample of limestone is heated strongly for 10 minutes. Water is then added drop by drop. What is observed? *[2]*

11 What is the common name for calcium hydroxide solution? *[1]*

12 What is the chemical name for rust? *[2]*

13 A piece of magnesium ribbon wrapped around an iron nail prevents it from rusting. Explain how the magnesium protects the iron from rusting. *[2]*

14 What colour change is observed when chlorine is bubbled through a solution of iron(II) sulphate? *[2]*

15 Compound **A** is a metal chloride. A flame test is carried out and the flame colour observed is blue-green.

a Explain how a flame test is carried out. *[5]*

b What is the identity of the metal chloride? *[1]*

11

Non-metals

Much of the non-metal chemistry section involves gases and it is important to know the following:

- the preparation of the gases
- the physical properties of the gases
- the chemical properties of the gases
- the test for the gases.

Preparation of the gases

You need to know:

- the reagents
- the labelled diagram of the apparatus

for the following gases:

- hydrogen
- carbon dioxide
- oxygen
- chlorine
- hydrogen chloride
- nitrogen by removing other gases from air.

Several factors contribute to the way in which a gas is collected.

Collecting insoluble gases

Figure 11.1 shows how gases which are **insoluble** in water or have a low solubility in water can be collected. This is called collection over water.

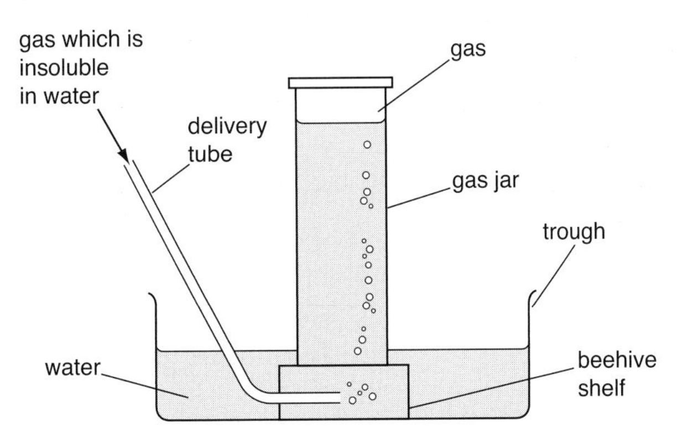

Figure 11.1 Apparatus used to collect a gas that is insoluble or has a low solubility in water

Collecting soluble gases

Gases that are **soluble** in water are collected by **displacement** of air, and the **density** of the gas compared to air must also be considered.

Soluble gases that are denser than air are collected by **downward delivery**. Those that are less dense than air must be collected by **upward delivery** (Figure 11.2a and b).

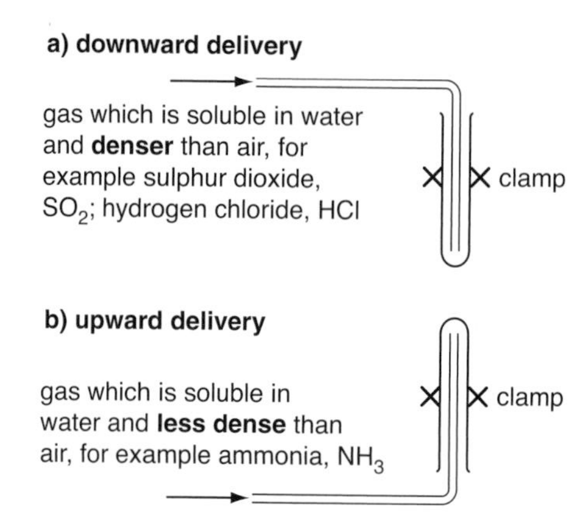

Figure 11.2 Apparatus used to collect a soluble gas that is **a)** denser or **b)** less dense than air

Hydrogen

Physical properties

There are four important physical properties of hydrogen. It is a colourless, odourless gas which is insoluble in water and less dense than air.

Chemical properties

1 A sample of hydrogen gas burns explosively.
  Observations: burns with a clean blue flame producing water vapour that may condense on glass
  Equation: $2H_2 + O_2 \rightarrow 2H_2O$
2 Hydrogen gas reacts with heated copper(II) oxide (page 110).
  Observations: black copper oxide changes to pink; condensation appears
  Equation: $CuO + H_2 \rightarrow Cu + H_2O$
  Apparatus: (see Figure 10.5 on page 110)
3 Hydrogen reacts with nitrogen forming ammonia in the Haber-Bosch process.
  Equation: $N_2 + 3H_2 \rightarrow 2NH_3$
  Industrial process: (see page 131)

Preparation of hydrogen

Hydrogen is prepared using zinc (or magnesium) and dilute hydrochloric acid (Figure 11.3).

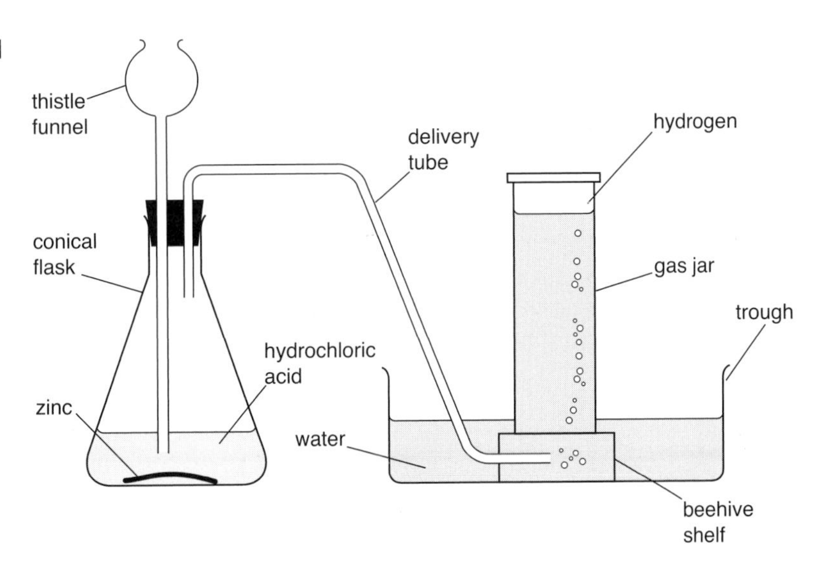

Figure 11.3 Apparatus used to prepare hydrogen

- Hydrogen is collected over water as it is **insoluble** in water.
- The reaction with magnesium is more vigorous and not recommended for a controlled preparation of hydrogen gas.
- The equations for the reactions are as follows:

$$Zn + 2HCl \rightarrow ZnCl_2 + H_2$$
$$(Mg + 2HCl \rightarrow MgCl_2 + H_2)$$

HINT: There are several important points to note when drawing this general gas apparatus:

- the thistle funnel must be *below* the level of the acid in the conical flask
- the delivery tube must not cut through the side of the trough
- the water level in the trough must be *above* the beehive shelf
- water must be present in the gas jar.

The marks in a diagram question are always awarded for the correct labels.

Test for hydrogen

Method

Apply a lighted splint to the gas.

Test result

Burns with a squeaky 'pop'.

The equation for this test is the combustion of hydrogen:

$$2H_2 + O_2 \rightarrow 2H_2O$$

Uses of hydrogen

Hydrogen is used:

- in meteorological (weather) balloons
- in rocket engines
- as a clean fuel.

Carbon

Carbon exists as the two distinct **allotropes**, diamond and graphite (page 29). The two forms have identical atoms but it is the way in which the atoms are bonded together that makes the forms different. Different forms of the same element in the same physical state are called allotropes.

Diamond

- Hardest naturally occurring substance.
- Very high melting point (approximately 3500 °C).
- Diamond-tipped tools are used for cutting glass/drilling/engraving.
- Each carbon atom is strongly bonded to four others in a **tetrahedral** arrangement.
- The strength and number of bonds accounts for the **very high melting point** as it takes a large amount of energy to break all the bonds.
- Does not conduct electricity.
- Structure – see page 29.

Graphite

- Only non-metal to conduct electricity.
- Very high melting point (approximately 3600 °C).
- Layered structure with weak forces of attraction between the layers. This means that the layers can slide over each other. This accounts for the flakiness of graphite and its uses in pencil leads and as a lubricant.
- Each carbon atom is strongly bonded to three others in a **hexagonal** arrangement.
- Has **delocalised** electrons between the layers that explain why graphite can conduct electricity.
- Structure – see page 29.
- The forces of attraction between the layers are weak and so can be broken easily hence allowing the layers to slide over each other. However the strong covalent bonds in the layers give graphite its **high melting point** as they require a large amount of energy to break them.

Buckminsterfullerene

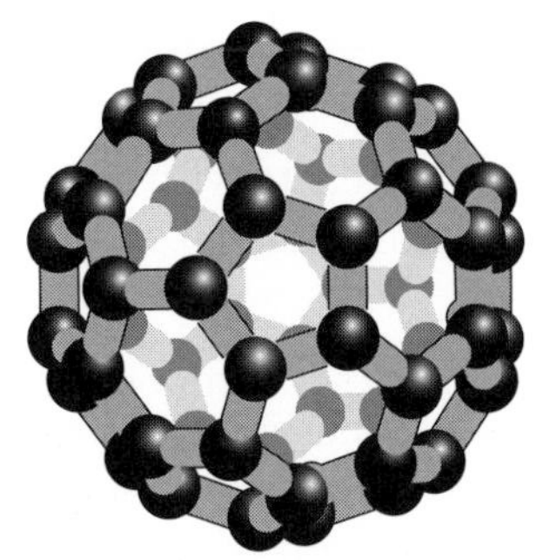

Figure 11.4 The structure of buckminsterfullerene

Carbon can also form a large molecule which is a covalently bonded 'ball' consisting of hexagons and pentagons of 60 carbon atoms (Figure 11.4). Its formula is C_{60} and its structure is like a football. It is called **buckminsterfullerene**. Buckminsterfullerene is soluble in organic solvents like toluene.

Each black sphere in the diagram represents a carbon atom and the grey lines represent covalent bonds.

Chemical properties of carbon

Carbon burns in an excess of oxygen.

Observations: it burns with a dirty orange flame, forming a colourless, odourless gas

Equation: $C + O_2 \rightarrow CO_2$

In a limited supply of oxygen, colourless carbon monoxide is formed.

Equation: $2C + O_2 \rightarrow 2CO$

Often in a limited supply of oxygen unburnt carbon appears as black **soot**. Carbon monoxide is extremely **toxic**.

In domestic fires, if there is an insufficient supply of oxygen to the fire carbon monoxide gas can be given off, which leads to the death of the occupants due to the toxic effects of the gas.

Chemical properties of carbon dioxide

1. Reaction of carbon dioxide with water.
   Carbon dioxide reacts with water to form the weak acid carbonic acid, H_2CO_3.
   Equation: $CO_2 + H_2O \rightarrow H_2CO_3$
   H_2CO_3 cannot be isolated from the solution and so is often simply written CO_2 (aq).
   Carbonic acid causes the acidity in fizzy drinks.
2. Reaction of carbon dioxide with burning magnesium.
   A piece of burning magnesium burns in a gas jar of carbon dioxide.
   Observations: a bright white light; a white solid is produced which is magnesium oxide and specks of black carbon
   Equation: $2Mg + CO_2 \rightarrow 2MgO + C$

3 Reaction of carbon dioxide with an alkali. Carbon dioxide is an acidic oxide and reacts with an alkali producing a salt and water (pages 45 and 62).
Equation: $CO_2 + 2NaOH \rightarrow Na_2CO_3 + H_2O$
In the above equation carbon dioxide reacts with sodium hydroxide solution, producing a colourless solution of the salt sodium carbonate and water.

Preparation of carbon dioxide

Carbon dioxide is prepared from calcium carbonate (marble chips) and hydrochloric acid using the apparatus shown in Figure 11.3 on page 127. The zinc is replaced with calcium carbonate (marble chips). Carbon dioxide is collected over water as it has a low solubility in water.

Equation: $CaCO_3 + 2HCl \rightarrow CaCl_2 + CO_2 + H_2O$

Test for carbon dioxide

Method
Bubble gas through limewater.

Test result
Colourless solution changes to 'milky'.
If carbon dioxide is bubbled through limewater until it is in excess, the colourless solution changes to milky due to a white precipitate and then the precipitate re-dissolves to form a colourless solution (see page 116).

Equations: $CO_2 + Ca(OH)_2 \rightarrow CaCO_3 + H_2O$
$Ca(OH)_2$: limewater; $CaCO_3$: white ppt

$CaCO_3 + CO_2 + H_2O \rightarrow Ca(HCO_3)_2$
$CaCO_3$: white ppt; $Ca(HCO_3)_2$: colourless solution

Uses of carbon dioxide

Carbon dioxide is used in:
- fire extinguishers
- carbonated drinks
- dry ice.

Nitrogen

Physical properties

Nitrogen is a colourless, odourless gas that is insoluble in water. It is a **diatomic** gas which means it consists of two atoms. It is written as N_2.

Preparation of nitrogen

Nitrogen is prepared in the laboratory by removing carbon dioxide, oxygen and water vapour from a sample of air.

- Carbon dioxide is removed by bubbling air through an alkaline solution such as sodium hydroxide:

  $$CO_2 + 2NaOH \rightarrow Na_2CO_3 + H_2O$$

- Oxygen is removed by passing the air over heated copper which reacts to form copper(II) oxide:

  $$2Cu + O_2 \rightarrow 2CuO$$

- Finally the water vapour in air can be removed by passing the remaining mixture of gases through concentrated sulphuric acid which acts as a **dehydrating agent**, removing the water.

The nitrogen produced is collected over water. It is **impure** as it contains Noble gases present in air, such as argon. These do not interfere with the reactions of nitrogen as they are **inert gases**.

Industrially nitrogen is manufactured by the **fractional distillation of liquid air**.

Chemical properties

1. Nitrogen is an unreactive gas but it does react with burning magnesium to form magnesium nitride, Mg_3N_2. Even magnesium burning in air forms about 10% magnesium nitride and 90% magnesium oxide.
2. Nitrogen and hydrogen react industrially to form ammonia. This is the **Haber-Bosch process**.
   - Nitrogen and hydrogen are mixed in a 1:3 ratio.
   - They are reacted at 450 °C, a pressure of 200 atm and with an iron catalyst.
     Equation: $N_2 + 3H_2 \rightarrow 2NH_3$
   - The gases are then cooled to condense the ammonia.
   - Unreacted nitrogen and hydrogen are recycled.
   - The Haber-Bosch process cannot be demonstrated in the laboratory due to the high pressure and temperature used, the specialised plant required and safety of staff and students. It would also be too expensive.

Uses of nitrogen

Nitrogen is used:

- as a coolant (as **liquid nitrogen**)
- in food packaging (nitrogen creates an **inert atmosphere** used to keep food fresh).

Ammonia

Physical properties

Ammonia is a colourless, pungent gas that is soluble in water and less dense than air.

Preparation of ammonia

Ammonia can be prepared in the laboratory by heating a solid ammonium compound with a solid alkali, for example sodium hydroxide or calcium hydroxide:

$$NaOH + NH_4Cl \rightarrow NaCl + H_2O + NH_3$$
$$Ca(OH)_2 + 2NH_4Cl \rightarrow CaCl_2 + 2H_2O + 2NH_3$$

The ammonia produced is collected by **upward delivery** (Figure 11.5).

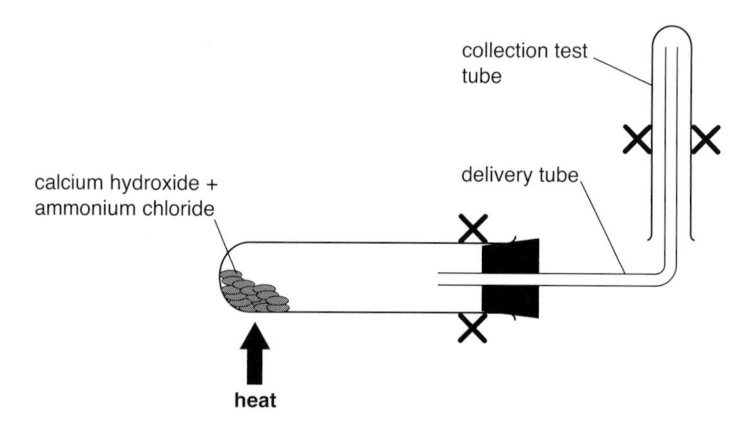

Figure 11.5 Apparatus used to prepare ammonia in the laboratory

Chemical properties

1 Ammonia is **basic** – it forms an alkaline solution due to the formation of ammonium hydroxide which produces **hydroxide ions** in water.
Equation: $NH_3 + H_2O \rightarrow NH_4OH$

2 Ammonia reacts with acids to form ammonium salts.

$$NH_3 + HCl \rightarrow NH_4Cl$$
$$2NH_3 + H_2SO_4 \rightarrow (NH_4)_2SO_4$$
$$NH_3 + HNO_3 \rightarrow NH_4NO_3$$

Test for ammonia

Method
Dip a glass rod in concentrated hydrochloric acid and apply to a sample of the gas.

Test result
If ammonia is present, a **white** '**smoke**' of ammonium chloride is observed.

Aqueous ammonia (ammonium hydroxide solution)

Aqueous ammonia can be used to test for the presence of metal ions in solution.

Copper(II) ions, Cu^{2+}

- Aqueous ammonia is added to a solution containing copper(II) ions.
- A **pale blue precipitate** (of copper(II) hydroxide, $Cu(OH)_2$) is formed:

$$Cu^{2+}(aq) + 2OH^-(aq) \rightarrow \underset{\text{pale blue ppt}}{Cu(OH)_2(s)}$$

- On addition of **excess aqueous ammonia** the **precipitate re-dissolves** to form a **dark blue solution**.

Iron(II) ions, Fe^{2+}

- Aqueous ammonia is added to a solution containing iron(II) ions.
- A **pale green precipitate** (of iron(II) hydroxide, $Fe(OH)_2$) is formed:

$$Fe^{2+}(aq) + 2OH^-(aq) \rightarrow \underset{\text{pale green ppt}}{Fe(OH)_2(s)}$$

- On addition of **excess aqueous ammonia** the **precipitate does not re-dissolve**.

Iron(III) ions, Fe^{3+}

- Aqueous ammonia is added to a solution containing iron(III) ions.
- A **red/brown precipitate** (of iron(III) hydroxide, $Fe(OH)_3$) is formed:

$$Fe^{3+}(aq) + 3OH^-(aq) \rightarrow \underset{\text{red-brown ppt}}{Fe(OH)_3(s)}$$

- On addition of **excess aqueous ammonia** the **precipitate does not re-dissolve**.

Aluminium ions, Al^{3+}

- Aqueous ammonia is added to a solution containing aluminium ions.
- A **white precipitate** (of aluminium hydroxide, $Al(OH)_3$) is formed:

$$Al^{3+}(aq) + 3OH^-(aq) \rightarrow \underset{\text{white ppt}}{Al(OH)_3(s)}$$

- On addition of **excess aqueous ammonia** the **precipitate does not re-dissolve**.

Zinc ions, Zn^{2+}

- Aqueous ammonia is added to a solution containing zinc ions.
- A **white precipitate** (of zinc hydroxide, $Zn(OH)_2$) is formed:

$$Zn^{2+}(aq) + 2OH^-(aq) \rightarrow \underset{\text{white ppt}}{Zn(OH)_2(s)}$$

- On addition of **excess aqueous ammonia** the **precipitate re-dissolves** to form a **colourless solution**.

Magnesium ions, Mg^{2+}

- Aqueous ammonia is added to a solution containing magnesium ions.
- A **white precipitate** (of magnesium hydroxide, $Mg(OH)_2$) is formed:

$$Mg^{2+}(aq) + 2OH^-(aq) \rightarrow \underset{\text{white ppt}}{Mg(OH)_2(s)}$$

- On addition of **excess aqueous ammonia** the **precipitate does not re-dissolve**.

Uses of ammonia

Ammonia is used in:
- fertilisers
- production of nitric acid
- nylon.

Nitric acid

Industrial manufacture

Nitric acid is manufactured from ammonia. There are three main stages of production.

Stage 1: Catalytic oxidation of ammonia
Ammonia and **air** are mixed and passed over a **platinum/rhodium** gauze at a temperature between **900 °C** and **1000 °C** and at a pressure of **2 atmospheres**.
The product is **nitrogen monoxide**, NO.
Equation: $4NH_3 + 5O_2 \rightarrow 4NO + 6H_2O$

Stage 2: Further oxidation of nitrogen monoxide
The gases are mixed with **more air** and nitrogen monoxide is converted to **nitrogen dioxide**.
Equation: $2NO + O_2 \rightarrow 2NO_2$

Stage 3: Formation of nitric acid
The gases are mixed with **more air** and passed up a tower of glass beads that have **water flowing down**.
Equation: $4NO_2 + O_2 + 2H_2O \rightarrow 4HNO_3$

Chemical properties

Nitric acid is a typical dilute acid that reacts with metal oxides, hydroxides and carbonates.

1 Reaction with metal oxides.
Equation: $MgO + 2HNO_3 \rightarrow Mg(NO_3)_2 + H_2O$
Observations: heat released, colourless solution produced

2 Reaction with metal hydroxides.
Equation: $Cu(OH)_2 + 2HNO_3 \rightarrow Cu(NO_3)_2 + H_2O$
Observations: heat released, blue solution produced

3 Reaction with metal carbonates.
Equation: $MgCO_3 + 2HNO_3 \rightarrow Mg(NO_3)_2 + CO_2 + H_2O$
Observations: heat released, colourless solution and bubbles of gas produced

HINT: You must be able to write equations for the reactions of any metal oxide, metal hydroxide or metal carbonate with nitric acid.

- Remember that the nitrate ion is NO_3^- and has a **valency** of 1.
- Nitric acid forms nitrate salts that are **all soluble**.
- All Group I and II nitrate salts are white and form colourless solutions.
- Aluminium and zinc nitrate are also white and form colourless solutions.

Uses of nitric acid

Nitric acid is used in:
- fertilisers
- explosives.

Environmental problems with nitrate fertilisers

Excessive use of **nitrate fertilisers** on soil leads to **leaching** of nitrates into soil water and finally into water courses. This leads to an environmental problem called **eutrophication** (page 59).

HINT: You need to be able to recognise the process of eutrophication and be able to explain it. Make sure you include the following points in your answer to a question on eutrophication:

- excess nitrates leads to excess growth of algae
- this leads to death of algae
- decomposers use up oxygen gas in water as they break down dead algae
- this leads to a lack of oxygen in the water
- and to death of fish.

Oxygen

Physical properties

Oxygen is a colourless, odourless gas that is only slightly soluble in water and is slightly denser than air.

Preparation of oxygen in the laboratory

- Oxygen is prepared from hydrogen peroxide and manganese dioxide using the apparatus shown in Figure 11.3 on page 127. The zinc is replaced with manganese dioxide and the hydrochloric acid is replaced with hydrogen peroxide.
- Manganese dioxide is also called manganese(IV) oxide and it is a **catalyst** used to speed up the decomposition of hydrogen peroxide (see also page 94):

$$2H_2O_2 \xrightarrow{MnO_2} 2H_2O + O_2$$

Test for oxygen

Method
A glowing splint is applied to a sample of the gas.

Test result
The glowing splint re-lights.

Uses of oxygen

Oxygen is used in:
- medicine
- welding
- rocket engines.

Chemical properties of oxygen

1 Reaction of oxygen with carbon.
Observations: black carbon burns with an orange sooty flame giving the colourless gas carbon dioxide, CO_2
Equation: $C + O_2 \rightarrow CO_2$
If there is a limited supply of oxygen, the combustion reaction produces carbon monoxide, CO.
Equation: $2C + O_2 \rightarrow 2CO$

2 Reaction of oxygen with sulphur.
Observations: yellow sulphur burns with a blue flame, giving the colourless pungent gas sulphur dioxide, SO_2
Equation: $S + O_2 \rightarrow SO_2$

3 Reaction of oxygen with calcium.
Observations: calcium burns (after significant heating) with a red flame, producing the white solid calcium oxide, CaO
Equation: $2Ca + O_2 \rightarrow 2CaO$

4 Reaction of oxygen with magnesium.

Observations: magnesium burns with a bright white flame, releasing heat and producing the white solid magnesium oxide, MgO

Equation: $2Mg + O_2 \rightarrow 2MgO$

5 Reaction of oxygen with hydrogen.

Observations: hydrogen burns with a clean blue flame, producing a colourless odourless gas which condenses to form a colourless liquid which is water, H_2O

Equation: $2H_2 + O_2 \rightarrow 2H_2O$

6 Reaction of oxygen with zinc.

Observations: zinc glows red on heating with oxygen, producing zinc oxide which is a yellow solid that changes to a white solid on cooling

Equation: $2Zn + O_2 \rightarrow 2ZnO$

7 Oxygen is vital in all **combustion** reactions, including cellular respiration.

- **Hydrocarbon fuels**, such as methane, CH_4, burn in a plentiful supply of oxygen (**complete combustion**) giving carbon dioxide and water and releasing energy in the form of heat. In a limited supply of oxygen (**incomplete combustion**), carbon monoxide and water are produced. Heat is also released but complete combustion produces more heat energy.

  Equations: $CH_4 + 2O_2 \rightarrow CO_2 + 2H_2O$ plentiful supply of oxygen

  $2CH_4 + 3O_2 \rightarrow 2CO + 4H_2O$ limited supply of oxygen

- **Glucose**, $C_6H_{12}O_6$, is oxidised to carbon dioxide and water by cellular reactions producing energy which is used in cells.

  Equation: $C_6H_{12}O_6 + 6O_2 \rightarrow 6CO_2 + 6H_2O$

Sulphur

Physical properties

Sulphur is a brittle yellow solid.

Allotropy of sulphur

Sulphur exists as three **allotropes**:

- rhombic sulphur
- monoclinic sulphur
- plastic sulphur.

Allotropes are different forms of the same element in the same physical state (page 128).

Chemical properties of sulphur

Yellow sulphur does not react with water or acids but burns with a blue flame in air/oxygen producing the colourless pungent gas sulphur dioxide, SO_2.

Equation: $S + O_2 \rightarrow SO_2$

Uses of sulphur

Sulphur is used in:

- vulcanising of rubber
- fungicide.

Physical properties of sulphur dioxide

Sulphur dioxide is a colourless, pungent gas that is soluble in water and denser than air.

Chemical properties of sulphur dioxide

1 Sulphur dioxide reacts with water to form the weak acid sulphurous acid, H_2SO_3.

Equation: $SO_2 + H_2O \rightarrow H_2SO_3$

Sulphurous acid forms salts called **sulphites**.
A sulphite ion is SO_3^{2-}.

- Many fossil fuels contain sulphur impurities that burn on combustion of the fuels to form sulphur dioxide, which is released into the atmosphere.
- Sulphur dioxide in the atmosphere reacts with rain water to form a weak acid called **sulphurous acid**. This causes **acid rain**.
- Acid rain has three main effects:
  - corrodes limestone buildings and statues
  - defoliates trees (loss of leaves)
  - pollutes lakes and rivers, killing fish.
- Acid rain can be prevented by:
  - removing sulphur from fossil fuels
  - using renewable energy resources
  - removing sulphur dioxide from factory/power station emissions
  - using cars fitted with catalytic converters
  - burning less fossil fuels.

2 Sulphur dioxide is an acidic oxide gas that reacts with alkalis, forming sulphite salts and water.

Equation: $SO_2 + 2NaOH \rightarrow Na_2SO_3 + H_2O$

sodium sulphite

Uses of sulphur dioxide

Sulphur dioxide is used:

- as a bleach
- in preservatives
- in fungicides.

Industrial manufacture of sulphuric acid (Contact process)

Sulphuric acid is manufactured from sulphur in the **Contact process**.
There are four main stages of production:

Stage 1: Combustion of sulphur
Sulphur is burned in air to form **sulphur dioxide**.
Equation: $S + O_2 \rightarrow SO_2$

Stage 2: Catalytic production of sulphur trioxide
The **sulphur dioxide** is mixed with more air at a temperature of 450 °C, a pressure of 2 atm and in the presence of the catalyst vanadium(V) oxide, V_2O_5.
Under these conditions the sulphur dioxide is converted to **sulphur trioxide**.
Equation: $2SO_2 + O_2 \rightarrow 2SO_3$

Stage 3: Absorption in concentrated sulphuric acid
The sulphur trioxide is **dissolved** in **concentrated sulphuric acid** to form **oleum**, $H_2S_2O_7$.
Equation: $SO_3 + H_2SO_4 \rightarrow H_2S_2O_7$

Stage 4: Dilution of oleum
The oleum is mixed with water to produce **sulphuric acid**.
Equation: $H_2S_2O_7 + H_2O \rightarrow 2H_2SO_4$

Sulphur trioxide is not mixed directly with water as the reaction is too **exothermic** and produces a **corrosive mist** that is difficult to contain.

Chemical properties of dilute sulphuric acid

Dilute sulphuric acid is a typical dilute acid that reacts with metals and metal oxides, hydroxides and carbonates.

1. Reaction with metals
   For example: $Mg + H_2SO_4 \rightarrow MgSO_4 + H_2$
2. Reaction with metal oxides
   For example: $MgO + H_2SO_4 \rightarrow MgSO_4 + H_2O$
3. Reaction with metal hydroxides
   For example: $Mg(OH)_2 + H_2SO_4 \rightarrow MgSO_4 + 2H_2O$
4. Reaction with metal carbonates
   For example: $MgCO_3 + H_2SO_4 \rightarrow MgSO_4 + CO_2 + H_2O$

HINT: You need to be able to write equations and state observations for the reactions of any metal, metal oxide, metal hydroxide or metal carbonate with sulphuric acid. Remember that the sulphate ion is SO_4^{2-} and has a valency of 2.

- Sulphuric acid forms sulphate salts most of which are soluble.
- All Group I and II sulphate salts are white and form colourless solutions.
- Aluminium and zinc sulphate are also white and form colourless solutions.
- Copper(II) sulphate forms a hydrated salt, $CuSO_4.5H_2O$ which is blue and on heating forms the anhydrous white solid, $CuSO_4$. The anhydrous solid is used as a chemical test for water – it changes from white to blue in the presence of water.

Uses of sulphuric acid

Sulphuric acid is used in:
- car batteries
- the manufacture of detergents
- the manufacture of fibres
- the manufacture of pigments.

Chemical properties of concentrated sulphuric acid

1 Reaction of concentrated sulphuric acid with sodium chloride.
Observations: bubbles of gas; fumes; heat released
Equation: $NaCl + H_2SO_4 \rightarrow NaHSO_4 + HCl$
This reaction is used to prepare hydrogen chloride gas in the laboratory. It is the first and only reaction where you will meet the hydrogen sulphate ion, HSO_4^-.
$NaHSO_4$ is sodium hydrogen sulphate.

2 Reaction of concentrated sulphuric acid with sugar (sucrose). Concentrated sulphuric acid **dehydrates** sugar to form carbon which is a black solid and water which is released as water vapour.
Observations: The sugar swells and rises in the container. The reaction is not immediate. Heat is released and there is a distinct caramel smell and a pungent odour. A black solid remains.
Equation: sugar $\xrightarrow{\text{conc } H_2SO_4}$ carbon + water

3 Reaction of concentrated sulphuric acid with hydrated copper(II) sulphate, $CuSO_4.5H_2O$.
Observations: the blue solid changes to white.
The concentrated sulphuric acid again behaves as a **dehydrating agent**, removing the **water of crystallisation** to form the white solid anhydrous copper(II) sulphate, $CuSO_4$.

4 Reaction of concentrated sulphuric acid with alcohols and organic acids forming **esters**.
The formation of the ester **ethyl ethanoate** is promoted using concentrated sulphuric acid, again as a **dehydrating agent** (page 157). Concentrated sulphuric acid acts most often as a dehydrating agent removing water.

Diluting concentrated sulphuric acid

On diluting concentrated sulphuric acid a large amount of heat is evolved so the acid must be diluted slowly with stirring to prevent too much heat being produced too quickly.

- Safety glasses and gloves should be worn.
- The concentrated acid should be added to a large volume of water, slowly with stirring. Water **must not** be added to the acid.

Group VII elements

- The Group VII elements (fluorine, chlorine, bromine, iodine and astatine) are known as the **halogens**.
- The halogens – in particular, chlorine, bromine and iodine – are a good example of a non-metal group in the Periodic Table.
- The chemistry of fluorine is not studied at this level as it is too reactive and astatine is too rare.

In general the halogen elements are **non-metallic** in most of their physical and chemical properties but iodine has a density greater than the metal titanium. Group VII elements are all powerful **oxidising agents**. Note that their oxidising power decreases as the group is descended.

Physical properties

Table 11.1 gives information about the physical properties of chlorine, bromine and iodine.

NOTE: Fluorine is a pale yellow gas; astatine is a black solid.

Table 11.1 The physical properties of chlorine, bromine and iodine

Physical property	Chlorine	Bromine	Iodine
melting point (°C)	−101	−7	113
boiling point (°C)	−34	59	184
colour	yellow-green	red-brown	dark grey
state at rtp*	gas	liquid	solid
odour	pungent	pungent	faintly irritating

*room temperature and pressure

- The halogens all exist as **diatomic** molecules, e.g. Cl_2, Br_2, I_2.
- Chlorine is easily liquefied by pressure and is stored and transported in the liquid state in steel containers.
- When chlorine and bromine dissolve in water, some of the halogen reacts with the water.
- Dissolved in organic solvents, iodine gives a purple solution but in ethanol and potassium iodide solution, in which it is very soluble, the solution is brown. Iodine is virtually insoluble in water.

Preparation of chlorine

The preparation of chlorine from **concentrated** hydrochloric acid requires the use of a strong **oxidising agent**. **Potassium permanganate** is used as the oxidising agent. It is represented by '[O]' in the equation for the reaction:

$$2HCl + [O] \rightarrow Cl_2 + H_2O$$

Chlorine gas is washed with water to remove any hydrogen chloride gas. It is then dried by passing it through **concentrated sulphuric acid**.

The gas is collected by **downward delivery** as it is denser than air. Potassium permanganate reacts so readily that the reaction proceeds in the cold (Figure 11.6).

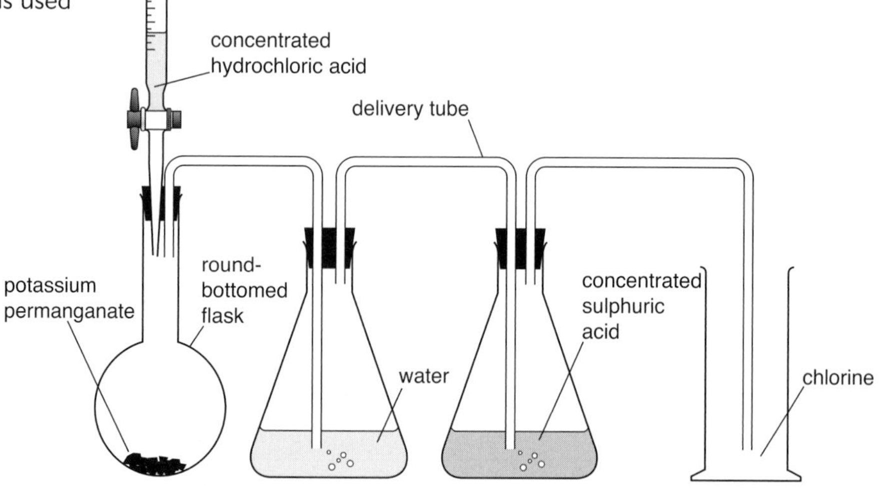

Figure 11.6 Apparatus used to prepare chlorine

Chemical properties of halogens

Most of the reactions of the halogens are based on their ability to **oxidise**.

1 Reaction of halogen with water.

Observations: yellow-green gas reacts with water and forms a pale greenish solution called **chlorine water** (mixture of two acids, HOCl and HCl)

Equation: $H_2O + Cl_2 \rightarrow HOCl + HCl$

- A reaction occurs between the chlorine and the water, giving an acidic solution.
- **Hypochlorous acid**, HOCl, is responsible for the fact that moist chlorine, unlike dry chlorine, is a powerful **bleaching agent**.
- Hypochlorous acid contains the hypochlorite ion, OCl^-.

2 Reaction of halogen with hydrogen.

Observations: hydrogen reacts explosively with chlorine, producing fumes of hydrogen chloride

Equation: $H_2 + Cl_2 \rightarrow 2HCl$

Hydrogen chloride fumes in air as it reacts readily with the water vapour in air. A solution of hydrogen chloride is called hydrochloric acid.

3 Reaction of halogen with metals (see page 118).

- Powdered metals often react spontaneously with chlorine, whereas coarsely divided metals may require heating.
- If the metal involved has more than one valency, chlorine will oxidise it to the highest valency. This is shown very clearly with iron:
  - when chlorine is passed over heated iron, the product is iron(III) chloride:

    $$2Fe + 3Cl_2 \rightarrow 2FeCl_3$$
  - iron(II) chloride can be prepared by the reaction of iron with hydrogen chloride
  - if chlorine is bubbled through a solution of an iron(II) salt or over the solid iron(II) compound, the chlorine oxidises the iron(II) to iron(III):

    Ionic equation: $2Fe^{2+} + Cl_2 \rightarrow 2Fe^{3+} + 2Cl^-$
  - the colour change observed is from a pale green solution of Fe^{2+} to the yellow solution of Fe^{3+}
  - this oxidation occurs slowly when a solution of iron(II) is left exposed to air but the change occurs immediately with chlorine:

    Half equations: $Fe^{2+} \rightarrow Fe^{3+} + e^-$ *oxidation*

    $Cl_2 + 2e^- \rightarrow 2Cl^-$ *reduction*

- as chlorine oxidises the iron(II) to iron(III), it becomes reduced to chloride ions. This is characteristic of an **oxidising agent**.

- Other metals react with chlorine to form chlorides as expected:

$$Zn + Cl_2 \rightarrow ZnCl_2$$
$$Mg + Cl_2 \rightarrow MgCl_2$$
$$2Na + Cl_2 \rightarrow 2NaCl$$
$$Cu + Cl_2 \rightarrow CuCl_2$$

4 Reaction of halogen with cold sodium hydroxide solution.
Chlorine reacts with cold sodium hydroxide solution to give a mixture of a chloride and a hypochlorite:

$$Cl_2 + 2NaOH \rightarrow NaCl + NaOCl + H_2O$$

This may be given as the ionic equation with the sodium ion, Na^+, left out as the spectator ion:

$$Cl_2 + 2OH^- \rightarrow Cl^- + OCl^- + H_2O$$

Sodium hypochlorite is the salt of **hypochlorous acid**.

5 Reaction of halogen with halides.
Each halogen displaces those below it in the Periodic Table from solution.

- Chlorine displacing bromide ions from solution

Observations: the yellow-green chlorine dissolves in the solution and the colour of the solution changes from colourless to brown

Equation: $Cl_2 + 2KBr \rightarrow Br_2 + 2KCl$

Ionic equation: $Cl_2 + 2Br^- \rightarrow Br_2 + 2Cl^-$

- Chlorine displacing iodide ions from solution

Observations: the yellow-green gas dissolves in the solution and the colour of the solution changes from colourless to brown

Equation: $Cl_2 + 2NaI \rightarrow I_2 + 2NaCl$

Ionic equation: $Cl_2 + 2I^- \rightarrow I_2 + 2Cl^-$

NOTE: Iodine in an aqueous solution of iodide ions is brown. Iodine shows no reaction with solutions of chloride and bromide ions and bromine shows no reaction with solutions of chloride ions.

Typical question

What is observed when chlorine reacts with a solution of potassium iodide? *[3]*

Answer

colourless [1] solution changes to brown [1]; heat released [1]

HINT: Questions on the chemistry of the halogens and chlorine in particular are generally not well answered. You should ensure that you learn all the non-metals section thoroughly and make sure you know the observations for *all* the reactions of chlorine.

Uses of chlorine and bromine

- Chlorine is used to manufacture bleach for cotton, linen and wood pulp, and in PVC manufacture and in water sterilisation.
- Bromine is used to test for **unsaturation** in hydrocarbons (alkenes) (page 151).

Hydrogen chloride

Hydrogen chloride is a colourless, toxic, pungent-smelling gas. It fumes in moist air as it reacts with the water vapour. It is highly soluble in water.

Preparation of hydrogen chloride

Hydrogen chloride is prepared by adding concentrated sulphuric acid to sodium chloride. Rock salt may be used instead of sodium chloride. Heating is not vital but it does speed up the reaction. The gas is denser than air and so is collected by **downward delivery** (upward displacement of air) (Figure 11.7).

The equation for this reaction is:

$$NaCl + H_2SO_4 \rightarrow NaHSO_4 + HCl$$

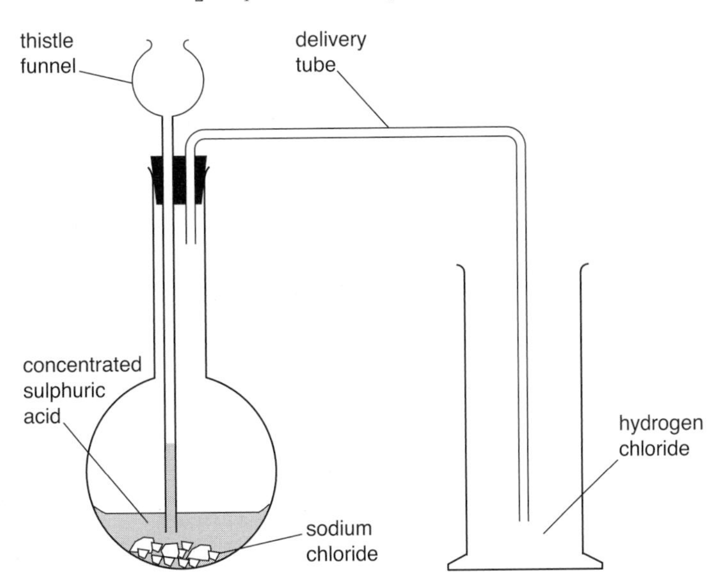

Figure 11.7 Apparatus used to prepare hydrogen chloride

Chemical properties of hydrogen chloride

1 Reaction of hydrogen chloride with water.
Hydrogen chloride dissolves in water creating an acidic solution called hydrochloric acid.
The reaction with water involves **dissociation** of the hydrogen chloride into a hydrogen ion and a chloride ion:

$$HCl\ (g) \xrightarrow{H_2O} H^+\ (aq) + Cl^-\ (aq)$$

The H^+ (aq) ion is what causes the solution to be acidic (see page 44).

2 Reaction of hydrogen chloride with ammonia.
Hydrogen chloride gas reacts with ammonia gas to form ammonium chloride.
Observations: dense white fumes (of ammonium chloride)
Equation: $NH_3 + HCl \rightarrow NH_4Cl$
This reaction is used as a test for hydrogen chloride gas and ammonia gas (page 132).

Hydrochloric acid

- Hydrochloric acid is a typical mineral acid and undergoes typical reactions with metals, metal oxides, metal hydroxides and metal carbonates.
- Almost all chlorides are soluble in water and so the salts produced are in solution.
- A metal ion in a compound that will not dissolve in water can be made to dissolve by reaction with a dilute acid, such as hydrochloric acid or nitric acid.

Chemical properties of hydrochloric acid

1 Reaction of hydrochloric acid with metals (page 105).
Most metals react with dilute hydrochloric acid to give the corresponding chloride and hydrogen gas.
Observations: bubbles of gas produced; heat released; metal disappears; colourless solution formed
Example equation: $Mg + 2HCl \rightarrow MgCl_2 + H_2$

NOTE: Copper and metals below copper in the reactivity series do not react with dilute hydrochloric acid.

2 Reaction of hydrochloric acid with metal oxides and hydroxides.
Metal oxides and hydroxides react with dilute hydrochloric acid, producing the corresponding chloride and water.
Observations: solid disappears; heat released; the colour of the solution formed depends on the metal ion present, for example a solution of copper chloride is blue but most are colourless solutions (see Appendix 1 on page 181)
Example equations: $CaO + 2HCl \rightarrow CaCl_2 + H_2O$
$Cu(OH)_2 + 2HCl \rightarrow CuCl_2 + H_2O$

3 Reaction of hydrochloric acid with metal carbonates.
All metal carbonates react with dilute hydrochloric acid, producing the corresponding chloride, carbon dioxide and water.

Observations: solid disappears; bubbles of gas produced; heat released; the colour of the solution formed depends on the metal ion present

Example equations:
$CaCO_3 + 2HCl \rightarrow CaCl_2 + CO_2 + H_2O$
$K_2CO_3 + 2HCl \rightarrow 2KCl + CO_2 + H_2O$

Noble gases

The Noble gases are **inert** and their uses are based on their lack of chemical reactivity.

Helium is used in balloons and in breathing mixtures for divers. Neon is used in lighting.

Revision questions

1 Name **two** chemicals needed to prepare each of the following gases:
- **a** oxygen [2]
- **b** carbon dioxide [2]
- **c** hydrogen [2]
- **d** hydrogen chloride [2]
- **e** ammonia [2]
- **f** chlorine [2]

2 State how you would carry out the test for the following gases and what you would observe if the gas was present.
- **a** hydrogen [2]
- **b** carbon dioxide [3]
- **c** oxygen [2]
- **d** hydrogen chloride [4]

3 Explain, using an equation, why a solution of ammonia is alkaline. [4]

4 State **two** uses of carbon dioxide. [2]

5 What is observed when a solution of ammonia is added slowly until it is in excess to a solution containing copper(II) ions? [4]

6 In the industrial production of nitric acid:
- **a** Name the starting materials. [2]
- **b** Name the catalyst used in the first stage. [1]
- **c** What operating temperature is used in the first stage? [1]

7 What name is given to the industrial process for the production of sulphuric acid? [1]

8 Describe how you would **safely** dilute a sample of concentrated sulphuric acid. [3]

9 What name is given to a solution of hydrogen chloride gas? [1]

10 Copy and complete the table below giving the colour and physical state of each halogen.

Halogen	State at room temperature and pressure	Colour	
chlorine			[2]
fluorine			[2]
bromine			[2]
iodine			[2]

11 Carbon dioxide reacts with burning magnesium.
- **a** Write a balanced symbol equation for the reaction. [3]
- **b** What is observed during this reaction? [3]

12 Write a balanced symbol equation for the production of ammonia in the Haber-Bosch process. [3]

13 What is observed when solid potassium carbonate reacts with dilute nitric acid? [3]

14 What is meant by the term 'allotrope'? [2]

15 State the names of the **three** allotropes of sulphur. [3]

12

Organic chemistry

- Fossil fuels and living things are based on the element **carbon**.
- Organic chemicals are obtained from **crude oil**.
- Chemicals obtained from crude oil are called **hydrocarbons**.
- Hydrocarbons consist of **carbon and hydrogen** atoms *only*.
- Oil spillages cause the following **environmental problems**:
  - destroy habitats
  - harm diving birds
  - eyesore on beaches/shorelines.

HINT: The definition of a hydrocarbon is a very common question. The word *'only'* is worth a mark and it is important to include it in your answer.

Fractional distillation of crude oil

Fractional distillation separates crude oil into simpler mixtures of hydrocarbons called **fractions**.

- Fractional distillation is carried out in a fractionating column (Figure 12.1).
- The crude oil enters at the bottom as a hot, gaseous mixture.
- The fractionating column has bubble caps that allow gases to move upwards.
- The temperature decreases *up* the column.
- As the gases move up the column, hydrocarbons condense when the temperature of the column is the same as their boiling point.

Figure 12.1 A fractionating column is used to separate crude oil into its fractions

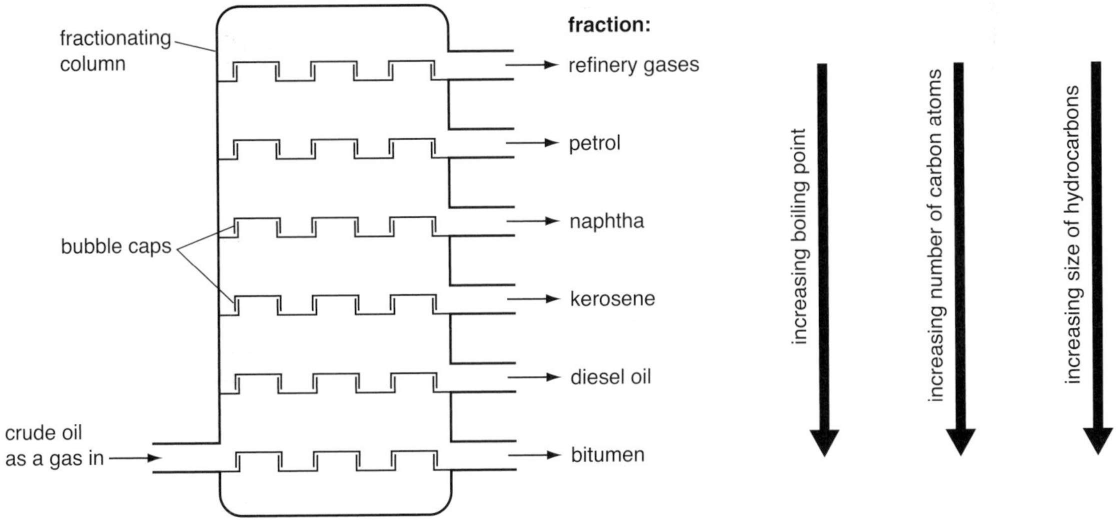

Table 12.1 shows the major fractions obtained from crude oil, in order of increasing size of molecules and increasing boiling point, with their uses.

Table 12.1 Major fractions obtained from crude oil

Name of fraction	Use
refinery gases	bottled gases
petrol	vehicle fuel
naphtha	chemicals and plastics
kerosene	aircraft fuel
diesel oil	large vehicle fuel
bitumen	road tar

Thermal cracking

Larger hydrocarbon molecules are not as useful as smaller ones. **Thermal cracking** is the breakdown of larger/longer hydrocarbon molecules into smaller/shorter ones which are more useful. The process uses heat in the absence of air.

Fossil fuels

- Fossil fuels are formed from dead plants and animals over millions of years under the action of heat and pressure.
- Examples of fossil fuels are natural gas, LPG, petrol, diesel, paraffin, candle wax, peat, lignite, coal, coke.
- Fossil fuels are **non-renewable resources** (they will eventually be used up).

Combustion of hydrocarbons

Combustion is the reaction of a fuel with oxygen, producing oxides and releasing heat.

Complete combustion is a fuel burning in a plentiful supply of oxygen/air, producing carbon dioxide and water and releasing heat.

Example equations: $CH_4 + 2O_2 \rightarrow CO_2 + 2H_2O$

$2C_2H_6 + 7O_2 \rightarrow 4CO_2 + 6H_2O$

Incomplete combustion is a fuel burning in a limited supply of oxygen/air, producing carbon monoxide and water and releasing heat.

Example equations: $C_2H_4 + 2O_2 \rightarrow 2CO + 2H_2O$

$2C_3H_8 + 7O_2 \rightarrow 6CO + 8H_2O$

Carbon monoxide is a poisonous gas (page 129).

HINT: The equations for combustion are often asked for in organic questions.

Balancing the equations

To balance a complete combustion equation:

1 The number of carbon atoms in the hydrocarbon is the same as the balancing number in front of the CO_2.

2 The number of hydrogen atoms in the hydrocarbon is divided by 2 to get the number in front of the H_2O.

3 Count the **total number of oxygen atoms** in the CO_2 (remember CO_2 has 2 per CO_2) and H_2O and divide by 2 to get the number in front of O_2.

4 If the number in front of O_2 has a half, for example $2\frac{1}{2}$, multiply all the balancing numbers by 2 to get whole numbers.

Testing products of combustion

The main products of combustion are carbon dioxide and water vapour (pages 59 and 116). The presence of these two products can be detected using two chemical tests.

- Bubble the gases through limewater – the presence of carbon dioxide changes the limewater from colourless to milky.
- Cool the gases and a colourless liquid will condense. Add this liquid to anhydrous copper(II) sulphate – the presence of water will change the anhydrous copper(II) sulphate from white to blue.

Carbon dioxide in the atmosphere

Carbon dioxide in the atmosphere leads to the **greenhouse effect**. Scientists think that increased levels of carbon dioxide are increasing the greenhouse effect which causes:

- global warming
- rising sea levels
- climate changes.

Homologous series

A **homologous series** is a family of organic molecules that have the same general formula, show similar chemical properties, show a gradation in their physical properties and differ by a 'CH_2' unit.

There are two hydrocarbon homologous series, called the **alkanes** and **alkenes**. Alkanes are relatively unreactive hydrocarbons. Alkenes are more reactive hydrocarbons.

Saturation and unsaturation

Alkenes have one C=C double bond per molecule – they are **unsaturated**. Alkanes have no C=C double bonds per molecule – they are **saturated**.

Testing for double bonds/unsaturation

Method
Add the substance to bromine water.

Test result

- For an unsaturated substance/any alk**e**ne: colour changes from red-brown to colourless
- For a saturated substance/any alk**a**ne: colour remains red-brown

Alkanes

The alkanes have the general formula C_nH_{2n+2}

Table 12.2, overleaf, names the first four alkanes in the homologous series, and gives their structural formula and their state at room temperature and pressure.

Table 12.2 The first four alkanes

n	Name	Molecular formula	Structural formula	State at room temperature and pressure
1	methane	CH_4	H H—C—H H	gas
2	ethane	C_2H_6	H H H—C—C—H H H	gas
3	propane	C_3H_8	H H H H—C—C—C—H H H H	gas
4	butane	C_4H_{10}	H H H H H—C—C—C—C—H H H H H	gas

Alkenes

The alkenes have the general formula C_nH_{2n}

Table 12.3 names the first two alkenes in the homologous series, and gives their structural formula and their state at room temperature and pressure.

Table 12.3 The first two alkenes

n	Name	Molecular formula	Structural formula	State at room temperature and pressure
2	ethene	C_2H_4	H H C=C H H	gas
3	propene	C_3H_6	H H H C C=C H H H	gas

Addition polymerisation

Polymerisation is the process of creating a long molecule from a small molecule which forms the repeating unit in the **polymer**.

- A polymer is a long molecule formed by bonding together small molecules into a chain.
- **Addition polymerisation** is the process of adding molecules together to form a polymer. The long molecule is the polymer.
- The simple molecule from which a polymer is formed is called the **monomer**.
- The monomer has a **double bond** between two carbon atoms.
- The monomer is an **alkene**.
- The polymer is shown as the monomer with only a single bond in a square bracket.
- '*n*' molecules of monomer must be at the beginning of the equation.
- The polymer structure has '*n*' after it to show that the polymer repeats '*n*' times.

General formula for addition polymer formation

Figure 12.2 shows the general formula for addition polymer formation.

Figure 12.2 General formula of an addition polymer

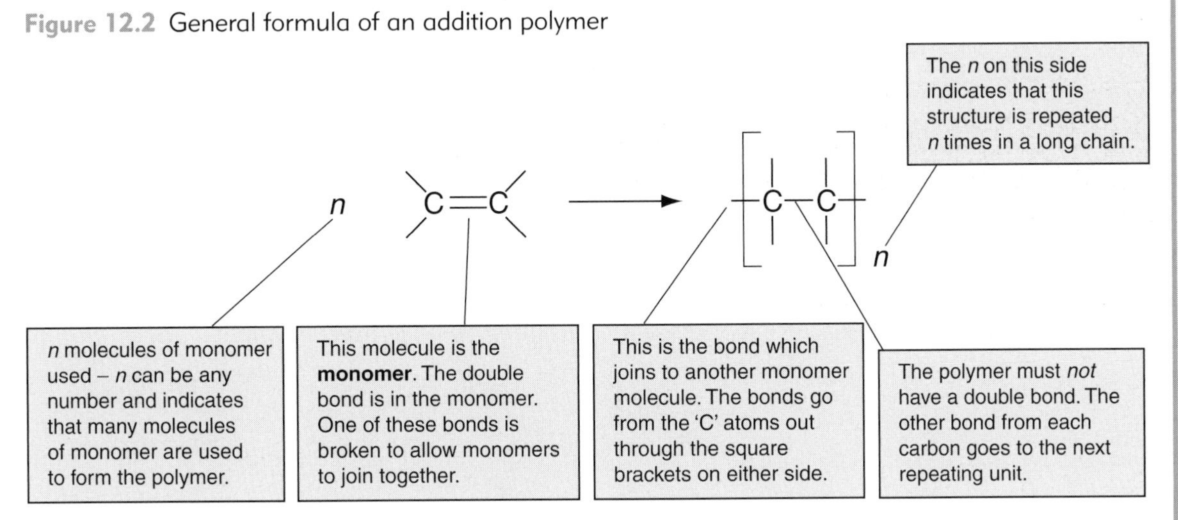

HINT: The equations for the formation of polymers are asked frequently in questions. The most common mistakes are forgetting to indicate the repeat in the polymer using '*n*', or putting a double bond in the structure of the polymer.

Common addition polymers

Polythene

Uses: plastic bags and bottles

$$n\ \mathrm{CH_2{=}CH_2} \longrightarrow \mathrm{-[CH_2-CH_2]_n-}$$

ethene
monomer

polythene
polymer

Poly vinyl chloride (PVC)

Uses: vinyl records, door and window frames, clothing

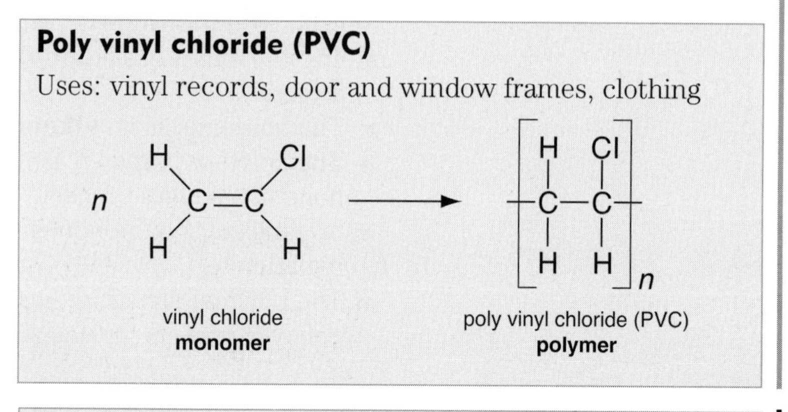

Polystyrene

Uses: packing material, insulation for houses

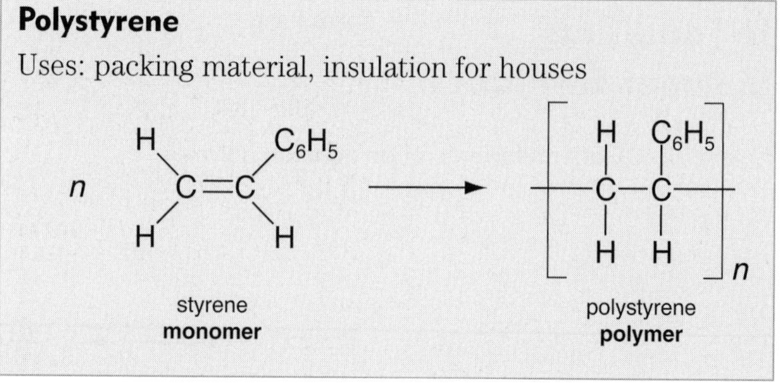

Polypropene

Uses: ropes, shampoo bottles, plastic crates

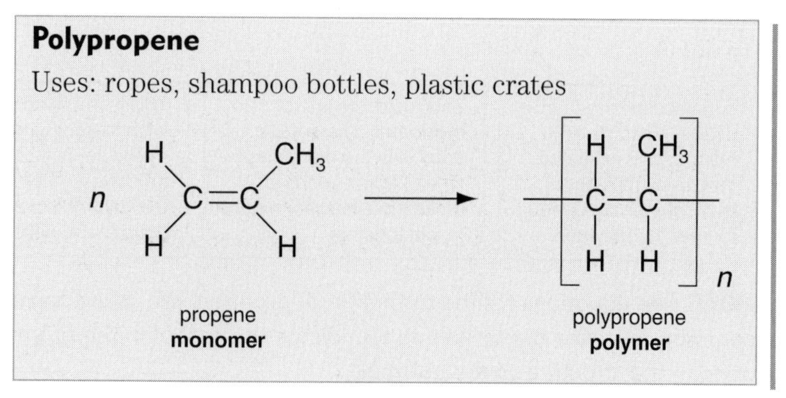

NOTE: Polypropene is not required for GCSE Chemistry.

Disposal of polymers

There are two methods of disposal of polymers:

- **landfill**
- **incineration**.

There are environmental problems associated with each method of disposal:

- landfill wastes land
- incineration produces polluting gases which are released into the atmosphere.

Polymers are increasingly being recycled, and biodegradable polymers have been developed and are now in use.

Other reactions of ethene

1 Combustion of ethene.

Observations: ethene burns with an orange flame, releasing heat

Equation: $C_2H_4 + 3O_2 \rightarrow 2CO_2 + 2H_2O$

2 Ethene reacts with steam, producing ethanol.

Equation: $C_2H_4 + H_2O \rightarrow C_2H_5OH$

Ethanol

- Molecular formula: C_2H_5OH
- Structural formula:

```
   H  H
   |  |
H—C—C—O—H
   |  |
   H  H
```

Typical question

Draw the structure of ethanol showing all the bonds. *[2]*

Answer

As shown above. *[2]*

HINT: The structures of ethanol and ethanoic acid often cause problems in answers to questions. The *full* structure *is* expected but often the '–O–H' at the end of the molecule is drawn as '–OH'. In ethanol and ethanoic acid this leaves out the bond between the 'O' and the 'H' atoms and is worth only 1 out of the 2 marks available. Make sure you get them both!

Reactions of ethanol

Combustion of ethanol.

Observations: burns with a clean blue flame, heat is released

Equation: $C_2H_5OH + 3O_2 \rightarrow 2CO_2 + 3H_2O$

Making ethanol

- Ethanol can be produced by the **fermentation** of sugars. Sugar solution is mixed with yeast in warm conditions in the absence of oxygen. The reaction produces carbon dioxide and ethanol.

  Equation: sugar → ethanol + carbon dioxide
- Ethanol can be made by reacting ethene with steam.

  Equation: $C_2H_4 + H_2O \rightarrow C_2H_5OH$

Uses of ethanol

Ethanol is used:
- in alcoholic drinks
- as a solvent.

Ethanoic acid

- Ethanoic acid is also called **acetic acid**. A dilute solution of ethanoic acid is **vinegar**.
- Molecular formula: CH_3COOH
- Structural formula:

```
     H     O
     |    //
  H—C—C
     |    \
     H     O—H
```

Reactions of ethanoic acid as an acid

1 Reaction of ethanoic acid with a metal (page 47).

Observations: bubbles of gas produced, heat released, metal disappears, colourless solution formed

General equation: ethanoic acid + metal → metal ethanoate + hydrogen

Example equation: $2CH_3COOH + Mg \rightarrow (CH_3COO)_2Mg + H_2$

2 Reaction of ethanoic acid with a metal oxide.

Observations: heat released, solid disappears, colourless solution formed

General equation: ethanoic acid + metal oxide → metal ethanoate + water

Example equation: $2CH_3COOH + CaO \rightarrow (CH_3COO)_2Ca + H_2O$

3 Reaction of ethanoic acid with a metal hydroxide.

Observations: heat released, solid disappears, colourless solution formed

General equation: ethanoic acid + metal hydroxide → metal ethanoate + water

Example equation: $2CH_3COOH + Mg(OH)_2 \rightarrow (CH_3COO)_2Mg + 2H_2O$

4 Reaction of ethanoic acid with a metal carbonate.

Observations: bubbles of gas, heat released, solid disappears, colourless solution formed

General equation:
ethanoic acid + metal carbonate → metal ethanoate + water + carbon dioxide

Example equation: $2CH_3COOH + Na_2CO_3 \rightarrow 2CH_3COONa + H_2O + CO_2$

Uses of ethanoic acid

Ethanoic acid is used as:

- a preservative
- flavouring.

Esterification (formation of an ester)

An **ester** is a compound formed between a **carboxylic acid** (such as ethanoic acid) and an **alcohol** (such as ethanol).

Observations: two layers are formed; sweet smell

Equation:

$$\underset{\text{ethanoic acid}}{CH_3COOH} + \underset{\text{ethanol}}{C_2H_5OH} \rightarrow \underset{\text{ethyl ethanoate}}{CH_3COOC_2H_5} + \underset{\text{water}}{H_2O}$$

Ethyl ethanoate

- Molecular formula: $CH_3COOC_2H_5$
- Structural formula:

```
    H         H   H
    |         |   |
H — C — C — O — C — C — H
    |   ||    |   |
    H   O     H   H
```

HINT: The structure of ethyl ethanoate is the most complicated you will meet at this level and that is why examiners ask for it! Make sure you learn the structure shown above.

Uses of ethyl ethanoate

Ethyl ethanoate is used as:

- a flavouring in foods
- a solvent.

Preparation of ethyl ethanoate

Mix equal volumes of ethanoic acid and ethanol in a boiling tube. Then add a few drops of concentrated sulphuric acid and warm gently.

Properties of organic compounds

- The first four alkanes (CH_4, C_2H_6, C_3H_8 and C_4H_{10}) are colourless gases.
- The first two alkenes (C_2H_4, C_3H_6) are colourless gases.
- All the polymers are white solids. Coloured dyes can be added to them to make coloured plastics.
- Ethanol is a colourless liquid with an alcohol-like odour.
- Ethanoic acid is a colourless liquid with a vinegar-like odour.
- Ethanol and ethanoic acid mix with water.
- Ethyl ethanoate is a colourless liquid with a fruity/solvent smell.
- Ethyl ethanoate does not mix with water (that's why the two layers are formed when it is produced from ethanol and ethanoic acid) (page 157).
- The higher the carbon content of an organic compound, the sootier (and more orange) its flame when it burns.
- The lower the carbon content of an organic compound, the less sooty (and more clean and blue) its flame when it burns.
- Ethanol burns with a cleaner, blue flame than the alkanes or alkenes.

NOTE: Very little of the organic section is Application, except perhaps writing equations for the reactions of alkanes, alkenes, ethanol, ethanoic acid and formation of ethyl ethanoate and the identification of an unknown organic compound (below).

Unknown organic compounds

You may be given information about a compound and have to identify it from its physical and chemical properties.

Example 1

A colourless organic compound has a melting point of −114 °C and a boiling point of 79 °C. When a sample is burned on a watch glass, a clean blue flame is observed. Suggest the identity of the compound.

The compound is a liquid (based on its melting and boiling points).

The clean blue flame indicates a low carbon content.

This would suggest **ethanol**.

Example 2

A gaseous hydrocarbon has a percentage carbon content by mass of 80%. Determine the formula of the hydrocarbon and name it. The RFM of the compound is 30 (see also page 74).

Hydrocarbons contain *only* carbon and hydrogen.

If 80% of the hydrocarbon is carbon, 20% is hydrogen.

Assuming 100 g of hydrocarbon:

mass of carbon = 80 g mass of hydrogen = 20 g

Calculate moles of the atoms of each element:

moles of carbon $= \frac{80}{12} = 6.67$ moles of hydrogen $= \frac{20}{1} = 20$

⇒ ratio of carbon to hydrogen = 6.67:20
⇒ simplest ratio of carbon to hydrogen = 1:3
⇒ simplest formula of hydrocarbon = CH_3
⇒ the simplest formula has an RFM of 15
⇒ the RFM of the simplest formula is half of the RFM of the molecule
⇒ the molecular formula is twice the empirical formula
⇒ the molecular formula is C_2H_6
⇒ the compound is **ethane**

Example 3

A gaseous hydrocarbon is bubbled through bromine water and the bromine water changes from red-brown to colourless. Identify the homologous series to which the hydrocarbon belongs.

Bromine water is decolourised when it reacts with C=C double bonds.

Alkenes contain C=C double bonds.

The unknown hydrocarbon is an **alkene**.

Revision questions

1 Write the molecular formula for each of the following molecules.

a ethane
b ethene
c butane
d methane *[4]*

2 What is observed when ethene is bubbled through bromine water? *[2]*

3 What is meant by the term 'hydrocarbon'? *[2]*

4 Name the following molecules.

a
```
    H  H  H
    |  |  |
H—C—C—C—H
    |  |  |
    H  H  H
```

b
```
    H  H
    |  |
H—C—C—O—H
    |  |
    H  H
```

c
```
H       H
 \     /
  C=C
 /     \
H       H
```

d
```
    H           H  H
    |           |  |
H—C—C—O—C—C—H
    |  ||       |  |
    H  O        H  H
```
[4]

5 Write a balanced symbol equation for the combustion of ethanol. *[3]*

6 What is produced when ethene reacts with steam? *[1]*

7 What process is used to separate the hydrocarbons in crude oil? *[2]*

8 What is thermal cracking? *[2]*

9 Write a balanced symbol equation for the reaction of magnesium with ethanoic acid. *[3]*

10 What is meant by the term 'homologous series'? *[4]*

11 Name the following polymers.

a
```
 ┌ H  H ┐
 | |  | |
—C——C—
 | |  | |
 └ H  H ┘n
```

b

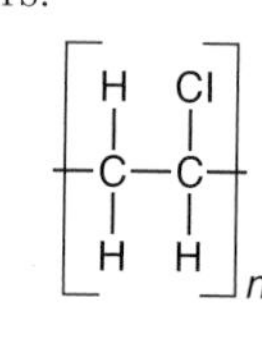

[2]

12 State **three** effects of increased carbon dioxide in the atmosphere. *[3]*

13 What are the products of incomplete combustion of ethane? *[2]*

14 What is a fossil fuel? *[3]*

15 State **one** use of kerosene. *[1]*

13

Chemical change

Classifying chemical reactions

There are **three** ways to classify a chemical reaction:

- energetics (pages 31–32)
- redox (pages 107–111)
- type of chemical reaction (pages 163–165).

NOTE: This chapter on chemical change is mostly a summary of types of reactions. The material consists of three major sections, all of which have been covered in previous chapters.

Energetics

Chemical reactions are either **exothermic** (give out heat to the surroundings) or **endothermic** (take in heat from the surroundings).

Table 13.1 shows important exothermic and endothermic reactions that you need to be able to state as examples.

Table 13.1 Examples of exothermic and endothermic reactions

Exothermic	Endothermic
neutralisation	photosynthesis
displacement	thermal decomposition
combustion	thermal cracking
hydration	electrolysis

All chemicals possess **internal energy** in their bonds. Energy is required to break all types of bonds and energy is released when all types of bonds are formed. This means that bond breaking is endothermic and bond making is exothermic.

HINT: In an energetics question your answer needs to be specific to the named reaction in the question. So your answer needs to follow this model: 'The energy required to break the bonds in the reactants (*name them*) is (*comparison – state less than for exothermic or more than for endothermic*) the energy released when bonds are made in the products (*name them*)'.

Typical questions

1 Methane burns in oxygen releasing energy to the surroundings. Explain why the combustion of methane is exothermic in terms of the energy of the bonds.

$$CH_4 + 2O_2 \rightarrow CO_2 + 2H_2O$$ *[5]*

2 The reaction of hydrogen with iodine to form hydrogen iodide is endothermic.

$$H_2 + I_2 \rightarrow 2HI$$

Explain, in terms of energy of bonds, why this reaction is endothermic. *[5]*

Answers

1 The energy required to break bonds *[1]* in the reactants which are methane and oxygen *[1]* is less *[1]* than the energy released when bonds are formed *[1]* in the products which are carbon dioxide and water *[1]*.

2 The energy required to break bonds *[1]* in the reactants which are hydrogen and iodine *[1]* is more *[1]* than the energy released when bonds are formed *[1]* in the product which is hydrogen iodide *[1]*.

Redox

A **redox reaction** is one in which **oxidation** and **reduction** occur simultaneously (page 107). Oxidation and reduction can be defined in one of the three ways given in Table 13.2.

Table 13.2 The three definitions of oxidation and reduction

Definition	Oxidation	Reduction
1	gain of oxygen	loss of oxygen
2	loss of hydrogen	gain of hydrogen
3	loss of electrons	gain of electrons

Many reactions can be simply described as oxidation or reduction in terms of the change in the oxygen or hydrogen content.

Other reactions can only be classified in terms of electrons lost or gained.

NOTE: Reduction is the reverse of oxidation.

HINT: The answer to a simple 'change in oxygen or hydrogen content' question is worth 2 or 3 marks. You should always give the following 3-mark answer to avoid losing any marks:

- species (*name it*) undergoes gain/loss in oxygen/hydrogen
- gain/loss of oxygen/hydrogen is oxidation/reduction (*as appropriate*)
- species has been oxidised/reduced.

Typical questions

3 Indicate whether the iron is being oxidised or reduced in the rusting of iron. *[3]*

4 Explain, in terms of electrons, how the reaction between zinc and copper(II) sulphate solution is a redox reaction. You may use ionic equations to help you answer this question.

zinc + copper(II) sulphate solution → copper + zinc sulphate solution *[7]*

Answers

3 iron has gained oxygen *[1]* gain of oxygen is oxidation *[1]* iron has been oxidised *[1]*

4 zinc atoms lose 2 electrons *or* $Zn \rightarrow Zn^{2+} + 2e^-$ *[2]*
oxidation is loss of electrons *or* zinc is oxidised *[1]*
copper ions gain 2 electrons *or* $Cu^{2+} + 2e^- \rightarrow Cu$ *[2]*
reduction is gain of electrons *or* copper ions are reduced *[1]*
redox is oxidation and reduction occurring simultaneously *[1]*

Type of chemical reaction

Apart from redox and energetics, there is another way of classifying chemical reactions which is by type of reaction.

The main types of reaction are:

- combustion
- displacement
- hydration
- neutralisation
- electrolysis
- thermal decomposition
- thermal cracking
- oxidation
- reduction.

Combustion

Combustion is the process of a fuel burning as it reacts with oxygen, forming oxides and releasing energy (page 150). All combustion reactions are therefore oxidation reactions and exothermic reactions. The main fuels used are **hydrocarbon fuels** which react with oxygen to form carbon dioxide and water and release energy:

fuel + oxygen → carbon dioxide + water (+ energy)

There are two major problems with our excessive use of fossil fuels (Table 13.3, overleaf).

Table 13.3 Effects of excessive use of fossil fuels

	Global warming	Acid rain
Cause	Carbon dioxide is a **greenhouse gas** which causes radiation from the Sun to be trapped near the Earth's surface. Increased levels of carbon dioxide are increasing this effect.	Fossil fuels contain sulphur impurities which are oxidised on combustion to sulphur dioxide: $S + O_2 \rightarrow SO_2$ The sulphur dioxide dissolves in atmospheric water to produce sulphurous acid (H_2SO_3), some of which is oxidised to sulphuric acid: $SO_2 + H_2O \rightarrow H_2SO_3$ $2H_2SO_3 + O_2 \rightarrow 2H_2SO_4$ These two acids create **acid rain**.
Effects	Increases the surface temperature causing: • climate changes • melting of polar ice caps • rise in seawater levels	Acid rain: • corrodes limestone statues and buildings • kills fish in rivers and lakes • defoliates trees

Acid rain is a national and an international responsibility as sulphur dioxide produced in one region or country can cause acid rain in another, due to the acid rain clouds being blown by winds.

Acid rain can be prevented by removing sulphur from fuels before use, treating emissions from power stations and factories to remove sulphur dioxide and by burning less fossil fuels.

Displacement

A **displacement reaction** is one where a more reactive element displaces a less reactive one from a compound (pages 107–109). It is a term mainly applied to a more reactive metal displacing a less reactive metal from a compound.

Example 1

Displacement reaction between zinc and copper sulphate solution.

Observations: blue solution fades to colourless; heat produced; brown/red solid produced

Equation: $Zn\ (s) + CuSO_4\ (aq) \rightarrow ZnSO_4\ (aq) + Cu\ (s)$

NOTE: The reaction is also a **redox reaction** due to the transfer of electrons as described previously. It is also **exothermic** as heat is produced – so all three classifications can be applied to this reaction.

Example 2

A displacement reaction can also occur between non-metals such as the halogens. A more reactive halogen displaces a less reactive one from its compound. Chlorine reacts with potassium iodide solution.

Observations: colourless solution changes to brown; heat produced

Equation: $2KI\ (aq) + Cl_2\ (aq) \rightarrow 2KCl\ (aq) + I_2\ (aq)$

The reaction is a displacement as the more reactive chlorine displaces the iodide ions to form chloride ions and iodine.

NOTE: The reaction is also a **redox reaction** due to the transfer of electrons:

Ionic equation: $2I^- + Cl_2 \rightarrow 2Cl^- + I_2$

Chlorine atoms gain electrons; gain of electrons is reduction; chlorine is *reduced*

$$Cl_2 + 2e^- \rightarrow 2Cl^-$$

Iodide ions lose electrons; loss of electrons is oxidation; iodide ions are *oxidised*

$$2I^- \rightarrow I_2 + 2e^-$$

This reaction is also a displacement reaction, a redox reaction and an exothermic reaction.

Hydration

- **Hydrated** salts contain **water of crystallisation** (page 68).
- Water of crystallisation is water which is chemically bonded in the crystal structure.
- When hydrated salts are heated they lose their water of crystallisation and form **anhydrous** salts.
- Loss of water of crystallisation involves breaking bonds in the crystals.
- The anhydrous salt remaining is a powder as water was providing the structure of the crystal.
- Removal of water of crystallisation is called **dehydration** and is an **endothermic** process as bonds are broken.
- Removal of water of crystallisation can be carried out using **concentrated sulphuric acid** and this reaction releases large amounts of heat as the water which is removed bonds with the concentrated acid.
- The most common hydrated salts used are:
  - hydrated copper(II) sulphate, $CuSO_4.5H_2O$ (blue)
  - hydrated cobalt(II) chloride, $CoCl_2.6H_2O$ (pink).

Hydration is the process where water is added to an anhydrous salt and bonds between the salt and water re-form releasing energy. The equations for these changes are written as follows:

$$CuSO_4 + 5H_2O \rightarrow CuSO_4.5H_2O$$

white → blue

$$CoCl_2 + 6H_2O \rightarrow CoCl_2.6H_2O$$

pale blue → pink

The addition of water to these two anhydrous salts is the basis of a **chemical** test for water. The colour changes indicate the presence of water.

Neutralisation

A **neutralisation reaction** between a **base** (or an alkali) and an **acid** produces a salt and water only (page 44). A base is any substance that neutralises an acid, producing a salt and water only. An alkali is a soluble base.

All neutralisation reactions are exothermic as they all have the following ionic equation in their reaction:

$$H^+(aq) + OH^-(aq) \rightarrow H_2O(l)$$

Example equation: $HCl + NaOH \rightarrow NaCl + H_2O$

Na^+ (aq) and Cl^- (aq) are **spectator ions** and the overall ionic equation is:

$$H^+ (aq) + OH^- (aq) \rightarrow H_2O(l)$$

which is an exothermic reaction

Electrolysis

Electrolysis is the process where a substance can be decomposed by a direct current of electricity (page 52).

In the industrial manufacture of aluminium, aluminium oxide dissolved in molten cryolite is electrolysed and the half equations at the carbon anode and carbon cathode are:

Cathode: $Al^{3+} + 3e^- \rightarrow Al$ aluminium ions are gaining 3 electrons, gain of electrons is reduction, aluminium ions are *reduced*

Anode: $2O^{2-} \rightarrow O_2 + 4e^-$ oxide ions are losing 2 electrons each, loss of electrons is oxidation, oxide ions are *oxidised*

The whole process involves the input of electricity which forces the ions to gain and lose electrons. The process of electrolysis is endothermic.

Thermal decomposition

Thermal decomposition is the process where a substance breaks down when heated (page 115). All thermal decompositions are endothermic.

Substances which undergo thermal decomposition are given below.

1 Some metal carbonates decompose when heated.
- Group I carbonates are stable to heat, except Li_2CO_3. For example, anhydrous Na_2CO_3 and K_2CO_3 do not decompose when heated.
- Group II carbonates and transition metal carbonates decompose when heated:

  metal carbonate → metal oxide + carbon dioxide

  Example equation: $CuCO_3 \rightarrow CuO + CO_2$

  Observations: $CuCO_3$ is a green solid which decomposes to the black CuO on heating

2 Hydrogen carbonates decompose when heated.
- Solid hydrogen carbonates of *only* sodium and potassium are known. Calcium hydrogen carbonate is only found in solution (page 117).
- All hydrogen carbonates decompose on heating:

metal hydrogen carbonate → metal carbonate + carbon dioxide + water

Example equation: $2NaHCO_3 \rightarrow Na_2CO_3 + CO_2 + H_2O$

Thermal cracking

Thermal cracking is another type of thermal decomposition that is applied to the breakdown of saturated hydrocarbons obtained from the fractional distillation of crude oil (page 149). The long chain hydrocarbons are broken down into smaller more useful ones. The process is carried out using *either* high temperatures and pressures *or* lower temperatures and pressure using a catalyst.

Example equation: $C_{15}H_{32} \rightarrow 2C_2H_4 + C_3H_6 + C_8H_{18}$

Some of the products must be **alkenes** (contain C=C double bonds).

Oxidation

As discussed on page 162, some reactions can simply be described as **oxidation** if they show:

- a gain of oxygen, for example,

  $S + O_2 \rightarrow SO_2$ sulphur is *oxidised*
- a loss of hydrogen, for example,

  $4NH_3 + 5O_2 \rightarrow 4NO + 6H_2O$ ammonia is *oxidised*
- a loss of electrons, for example,

  $2O^{2-} \rightarrow O_2 + 4e^-$ oxide ions are *oxidised*

Common oxidation reactions

1 Rusting

The most common example of an oxidation reaction is the rusting of iron in which iron metal reacts with oxygen in moist air and forms **hydrated iron(III) oxide** which is the chemical name for **rust** (page 119).

$$Fe + 1\tfrac{1}{2}O_2 + xH_2O \rightarrow Fe_2O_3.xH_2O$$

hydrated iron(III) oxide ('rust')

The value of x (the degree of hydration of rust) depends on the water content of the air in which the rust forms.

Rust can be prevented using the following methods:

- oiling
- greasing
- coating in plastic
- coating in a less reactive metal
- sacrificial protection.

Sacrificial protection (page 120) is when a more reactive metal is placed in contact with the iron. The more reactive metal reacts first instead of the iron, hence rusting is prevented. The most commonly used sacrificial metals are magnesium and zinc. For example, magnesium blocks are attached to the metal hulls of ships to prevent rusting.

Typical question

Explain why rusting is described as an oxidation reaction. *[2]*

Answer

Iron gains oxygen [1] and gain of oxygen is oxidation [1]
or iron loses electrons [1] and loss of electrons is oxidation [1]

HINT: The most common problem with students' answers to this question is that they are not specific to the reaction. Many answers state 'gain of oxygen is oxidation' without relating it to the iron.

2 Combustion

The **combustion** of fuels produces energy and the elements in the fuel (usually hydrogen and carbon) are oxidised to water and **carbon dioxide** (if the combustion is complete).

If the supply of oxygen is limited, carbon is only oxidised to **carbon monoxide** (this is incomplete combustion).

3 Respiration

Respiration is the cellular process whereby glucose is oxidised to carbon dioxide and water releasing energy which can be used by cells in the body.

4 Elements with oxygen
Magnesium is a grey metal that burns with a bright white light, releasing heat and forming a white powder.

$2Mg + O_2 \rightarrow 2MgO$

Sulphur is a yellow powder which burns in oxygen with a blue flame, producing heat and a colourless, pungent, choking gas which is sulphur dioxide.

$S + O_2 \rightarrow SO_2$

Reduction

As discussed on page 162, some reactions can simply be described as **reduction** if they show:

- a loss of oxygen, for example,

  $CuO + H_2 \rightarrow Cu + H_2O$

  CuO is *reduced*
- a gain of hydrogen, for example,

  $N_2 + 3H_2 \rightarrow 2NH_3$

  N_2 is *reduced*
- a gain of electrons, for example,

  $Al^{3+} + 3e^- \rightarrow Al$

  Al^{3+} ions are *reduced*

Reducing copper oxide
Metal oxides, such as copper oxide, can be reduced using hydrogen gas. The hydrogen gas is passed over the heated metal oxide (Figure 13.1).

Observations: black copper oxide changes to pink; condensation on inside of tube

Equation: $CuO + H_2 \rightarrow Cu + H_2O$

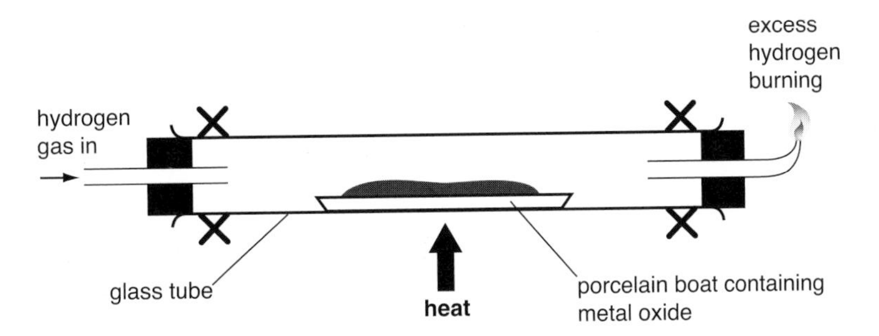

Figure 13.1 Apparatus used in the reduction of a metal oxide by hydrogen

Revision questions

1 What do the following terms mean?
- **a** exothermic [2]
- **b** endothermic [2]

2 Indicate whether the following reactions are exothermic or endothermic.
- **a** thermal decomposition of $CaCO_3$
- **b** neutralisation of sodium hydroxide with hydrochloric acid
- **c** hydration of anhydrous copper sulphate
- **d** complete combustion of methane [4]

3 Which combustion gas causes the greenhouse effect? [1]

4 Name **one** metal that can be used to prevent rusting by sacrificial protection. [1]

5 Explain, using **two** balanced symbol equations, how burning fossil fuels that contain sulphur impurities leads to the formation of acid rain. [4]

6 What is meant by the following terms?
- **a** combustion [3]
- **b** rust [2]
- **c** redox [1]

7 Write an equation to represent the reduction of aluminium ions in the extraction of aluminium by electrolysis. [3]

8 Write a balanced symbol equation for the thermal decomposition of copper(II) carbonate and state what you would observe. [4]

9 When zinc reacts with copper(II) sulphate solution, a displacement reaction occurs. The zinc is converted to zinc ions and the copper ions are converted to copper atoms.
- **a** Write an equation for the conversion of zinc atoms to zinc ions. [2]
- **b** Write an equation for the conversion of copper ions to copper atoms. [2]
- **c** State which of the above equations you have written is oxidation. [1]

10 Name **one** metal carbonate which cannot be decomposed by heating. [1]

11 State **three** effects of acid rain. [3]

12 In the Haber-Bosch process ammonia is formed from nitrogen and hydrogen according to the equation:

$$N_2 + 3H_2 \rightarrow 2NH_3$$

Explain why nitrogen is described as being reduced in this reaction. [2]

13 What would you observe when sulphur burns in air? [3]

14 Explain why controlling acid rain requires international co-operation. [2]

15 What **two** conditions are necessary for rusting to occur? [2]

14

Rocks, materials and radioactivity

Economics of industrial processes

For industrial manufacturing processes, the cost of manufacturing a chemical depends on the following.

- **Chemical** and **energy** factors:
  - the price, availability and method of transport of the raw materials
  - the construction of containers for reactions and storage
  - the energy used in the process (heat and electrical)
  - the costs of the pressure needed for the reaction to proceed and the vessels needed to contain the pressure.
- **Social** and **environmental** factors:
  - providing employment
  - preventing pollution of the environment with raw materials, products and waste from the process.

Chemical and energy factors

Table 14.1 gives the chemical and energy factors that affect industrial manufacturing processes and also gives the page numbers where details of the processes are given.

Table 14.1

Industrial process	Raw materials	Energy used/ pressure applied	Details of process
aluminium extraction	bauxite, cryolite, graphite electrodes	900–1000 °C, electricity	see page 56
iron extraction in the blast furnace	haematite, coke, limestone, hot air	heat (for hot air)	see page 120
thermal decomposition of limestone in a limekiln	limestone, coke, hot air	heat (for hot air)	see page 115
Haber-Bosch process	nitrogen, hydrogen, iron (catalyst)	450 °C, 250 atm	see page 131

(*Table 14.1 continues overleaf*)

(Table 14.1 continued)

Industrial process	Raw materials	Energy used/ pressure applied	Details of process
refining of copper	impure copper, pure copper, copper sulphate solution	electricity	see page 57
chloralkali industry	seawater, titanium	electricity	see page 53
nitric acid production	ammonia, air, water, platinum/rhodium (catalyst)	900–1000 °C, 2 atm	see page 134
sulphuric acid production (Contact process)	sulphur, air, vanadium(V) oxide catalyst, water, sulphuric acid	450 °C, 2 atm	see page 139

Social and environmental factors

Socially the industrial chemical companies provide employment. Environmental concerns are usually to do with pollution and use of land. For example, a limestone quarry provides employment in a region, improves roads, provides landfill sites and a local resource for use in agriculture and in road building.

Tables 14.2 and 14.3 give the major types of pollution and the concerns about the use of land.

Table 14.2 Types of pollution and the causes

Pollution type	Description and examples
air	gases (SO_2 from Contact process), dust (limestone from limestone quarrying)
thermal	hot water released by power stations into rivers and lakes decreases solubility of oxygen and kills fish
water	dissolved chemicals released into water, for example eutrophication from excessive use of nitrate fertilisers (page 59) oil pollution killing water birds and fish
noise	noise from machinery and traffic
land	plastic wastes pollute land as they are non-biodegradable

Table 14.3 Land use concerns

Concern	Description and examples
destruction of habitat	quarrying, peat cutting and building an industrial chemical plant, hedgerows cleared which are the habitat of many animals and birds
eyesore	construction of industrial plants, peat cutting and limestone quarrying create an eyesore on the landscape
subsidence	solution mining of salt removes underground deposits, destabilising the land above and causing landslides

Raw materials

The five major sources of raw materials and examples are shown in Table 14.4.

Table 14.4 Sources of raw materials and examples of the materials

Source of raw material	Examples of useful materials
the Earth	lime (from limestone), salt (from rock salt), aluminium (from bauxite), iron (from haematite)
sea	water, sodium chloride (salt)
air	oxygen, nitrogen, argon, other Noble gases
crude oil	plastics, petrol, diesel, bitumen
living things (plants and animals)	cotton, wool, wood, rubber

Natural and synthetic materials

Natural materials are those derived directly from raw materials. **Synthetic** materials are also called **man-made** materials. Raw materials are put through a manufacturing process to make a synthetic material. The following list gives examples of natural materials and synthetic materials.

- Natural materials: gold, limestone, rubber, wood
- Synthetic materials: plastics, iron, aluminium

NOTE: Synthetic materials are not required for GCSE Chemistry.

Chemical property related to its use

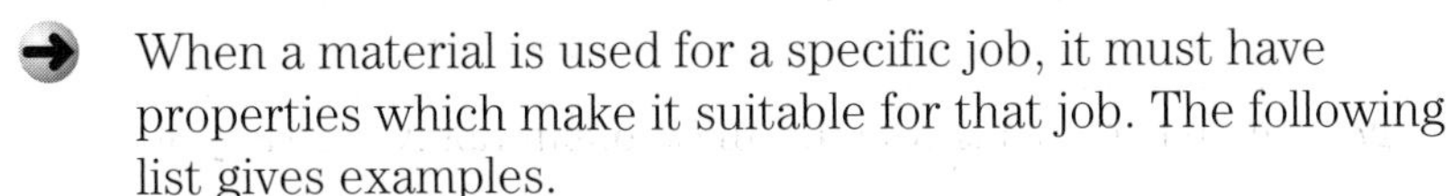

When a material is used for a specific job, it must have properties which make it suitable for that job. The following list gives examples.

- Copper is used in electrical wiring as it conducts electricity and is ductile.

- Glass is used for windows as it is transparent and can be formed into sheets.
- Aluminium is used in saucepans as it conducts heat and is malleable.

Synthetic materials

NOTE: Synthetic materials are not required for GCSE Chemistry.

Table 14.5 gives the five different groups of synthetic (man-made) materials and examples of the materials and their properties.

Table 14.5 Groups of synthetic materials and examples

Group of materials	Examples	Properties of materials
metals	copper, aluminium, iron, lead, zinc	conduct electricity, conduct heat, malleable, ductile, high melting point
ceramics	pottery, tiles	brittle, high melting point
glass	soda glass, Pyrex (heat-resistant glass)	brittle, high melting point, transparent
fibres	nylon	flexible, low melting point
plastics	**thermosetting** plastics such as Bakelite, Melamine, epoxy resins	brittle, high melting point, electrical insulators
	thermosoftening plastics such as polythene, PVC, polystyrene	flexible, can be moulded, low melting point, electrical insulators

Thermosetting and thermosoftening plastics

- Thermosetting plastics can be moulded. They are heated to set the shape but they cannot be melted and remoulded.
- Thermosoftening plastics can be moulded and heated to melt them and remould them.
- Thermosetting plastics contain cross-links between polymer chains when heated; thermosoftening plastics do not contain cross-links.

Composite materials

NOTE: Composite materials are not required for GCSE Chemistry.

A **composite** material is one which combines the properties of more than one material to produce a more useful material for a particular purpose. Examples of common composite materials are given in Table 14.6.

HINT: The most common error when asked for the definition of a composite material is to leave out the term 'properties'.

Table 14.6 Composite materials and examples of their uses

Composite material	Made from	Examples of uses
glass fibre (glass reinforced plastic)	fibres of glass and plastic fibres	loft insulation, boats and car bodies
reinforced concrete	steel rods inside concrete beams	construction
reinforced glass	glass with steel wires	security glass
bone	calcium phosphate and protein	skeleton

A composite material combines the best properties of each material. Reinforced glass combines the transparency of glass with the strength of steel. Bone combines the strength of calcium phosphate with the flexibility of protein.

NOTE: You need to be able to describe composite materials and their uses. It is important that you can describe the specific properties of each material.

Hazard symbols

Hazard symbols are used in the labelling of chemicals to indicate a risk or risks associated with them (Table 14.7). The symbols are recognised internationally.

Table 14.7 Hazard symbols and their meanings

Hazard symbol	Meaning	Hazard symbol	Meaning
i	irritant		toxic
h	harmful		flammable
	corrosive		explosive

Radioactivity

NOTE: This topic (pages 176 to 180) is required only for Double Award Science, not for GCSE Chemistry.

Radiation is something which is sent out or radiated from an object.

Isotopes

Isotopes of an element are atoms which contain different numbers of neutrons and hence have a different mass. If a nucleus contains too many neutrons, it may be unstable and break down, releasing energy as **ionising radiation**. Many isotopes are stable as the number of neutrons does not cause them to break down.

A common isotope of carbon can be written as ^{12}C or carbon-12.

Examples of stable and unstable isotopes

- Stable isotopes: carbon-12 (^{12}C), gold-197 (^{197}Au)
- Unstable isotopes: carbon-14 (^{14}C), uranium-238 (^{238}U)

When the nucleus of an atom of an unstable isotope breaks down, it gives out radiation. An element or isotope which gives out radiation is said to be **radioactive**. Carbon-14 and uranium-238 are examples of **radioactive isotopes**.

Types of radioactive emissions

There are three different types of radioactive emissions from unstable isotopes:

- alpha particles (α)
- beta particles (β)
- gamma radiation (γ).

All three types of radiation are **ionising** (can produce ions) but they differ in their properties. They are discussed individually below.

Alpha (α) particles

- **alpha (α) radiation** is made up of fast moving helium, $^{4}_{2}He$, nuclei.
- Each **alpha particle** is composed of two protons and two neutrons.
- An alpha particle has a positive charge and hence is deflected by an electric field towards the negative electrode.
- Alpha particles are also deflected by a magnetic field.
- Alpha particles are quite easily stopped by thin materials, such as a few sheets of paper. Even air stops them.
- Alpha particles are described as having a low penetrating power.

Beta (β) particles

- **Beta (β) radiation** is made up of electrons ($_{-1}^{0}e$) moving at high speed.
- **Beta particles** have a negative charge.
- Like alpha particles, beta particles are deflected by an electric field.
- Beta particles are deflected towards the positive electrode (in the opposite direction to alpha particles).
- Beta particles are also deflected in a magnetic field (again in the opposite direction to alpha particles).
- Beta particles are not stopped as easily as alpha particles.
- It requires a few centimetres of aluminium to stop beta particles and they travel several metres in air.
- Beta particles are described as having a moderate penetrating power.

Gamma (γ) radiation

- **Gamma (γ) radiation** is *not* a stream of particles. It is a form of **electromagnetic radiation**, i.e. a high-energy wave.
- This is why gamma radiation is sometimes referred to as gamma *rays*.
- Gamma radiation does not carry a charge and so is not deflected by either a magnetic field or an electric field – it passes straight through both.
- Gamma radiation is a very high-energy radiation.
- It can travel a very long way in air and can even pass through several centimetres of lead or even thicker pieces of concrete. It is only stopped by very thick concrete.
- Gamma radiation is said to have a high penetrating power.
- Gamma radiation can damage DNA and cause cells to divide uncontrollably, causing cancer.

Radiation penetrating power

Figure 14.1 summarises the penetrating powers of the three types of radiation.

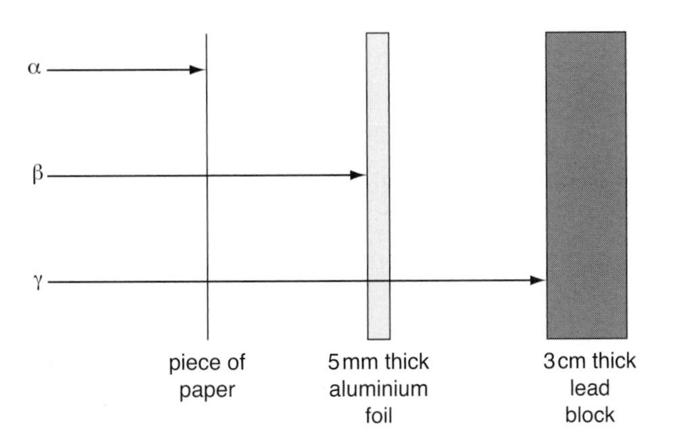

Figure 14.1 Comparing the penetrating powers of the three types of radiation

Radioactive decay

Radioactive decay occurs when a nucleus of an unstable isotope of one element decays to produce a more stable isotope. The more stable isotope is usually a different element.

Example 1

Carbon-14 has 6 protons and 8 neutrons in the nucleus of its atoms. This is unstable. The ^{14}C nucleus decays and emits a β particle.

One of the neutrons breaks down to a proton and an electron and the electron is emitted as β radiation.

This leaves the nucleus with 7 protons and 7 neutrons.

The nucleus now has 7 protons in its nucleus and so is the nucleus of a nitrogen atom.

It has 7 neutrons as well, so the mass of the nucleus is 7 + 7 = 14.

Carbon-14 decays, emitting a β particle, and producing nitrogen-14:

$$^{14}_{6}C \rightarrow ^{0}_{-1}e + ^{14}_{7}N + \text{energy}$$

Example 2

Uranium-238 is unstable. It decays and emits an α particle.

The α particle is written as $^{4}_{2}He$.

The nucleus simply loses 2 protons and 2 neutrons so the atomic number goes down by 2 and the mass number goes down by 4.

Uranium-238 decays to produce thorium-234:

$$^{238}_{92}U \rightarrow ^{4}_{2}He + ^{234}_{90}Th + \text{energy}$$

Detecting radiation

Radiation blackens photographic film and can also be detected by using a **Geiger-Müller tube** (sometimes called a Geiger counter). Radiation causes ions to be produced in the Geiger counter. The counter clicks for every ion detected.

Radiation can be measured in counts per second (cps). This is the same as ionisations per second. The higher the cps reading, the more radioactivity is being emitted.

Half-life of radioactive isotopes

For any particular radioactive isotope, the time taken for half a given number of radioactive atoms to decay is always the same and this time is called the **half-life**.

The half-life is often written $t_{\frac{1}{2}}$. The more unstable the nuclei, the more quickly they decay and the shorter the half-life. The longer the half-life, the more stable the isotope. Table 14.8 gives examples of half-lives.

Table 14.8 Half-lives of two radioactive isotopes

Radioactive isotope	Half-life ($t_{\frac{1}{2}}$)
uranium-238 $^{238}_{92}U$	4.5 billion years
carbon-14 $^{14}_{6}C$	5700 years

HINT: It is important to realise that it takes the same time for 10 g of a radioactive substance to decay to 5 g as it does for 5 g of the same substance to decay to 2.5 g. The half-life can also be determined as the time when the number of counts per second (cps) decreases by half.

Using a radioactive decay curve

Figure 14.2 shows a typical **decay curve** for a radioactive substance.

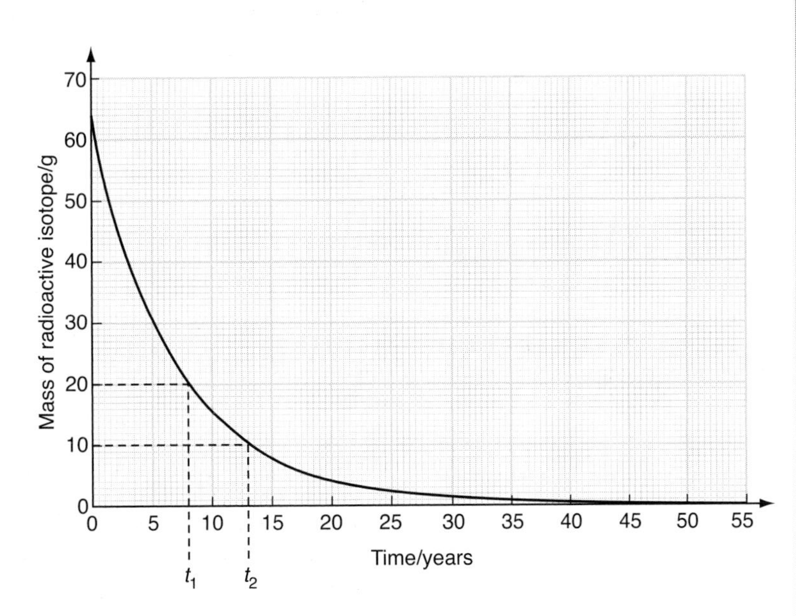

Figure 14.2 Graph to show the decay of a radioactive substance

The graph can be used to calculate the half-life of the radioactive substance – see overleaf.

Method of calculating half-life from a decay curve

1. Take a value on the mass axis (y axis) and read across to find the time in years on the x axis (t_1).
2. Take half the value on the mass axis and read across again to find the time in years (t_2).
3. Take one time away from the other to calculate the half-life ($t_{\frac{1}{2}} = t_2 - t_1$).

Example 3

From the graph in Figure 14.2:
mass = 20 g $t_1 = 8$ years
For mass = 10 g $t_2 = 13$ years
$t_{\frac{1}{2}} = 13 - 8 = 5$ years.

NOTE: This will work for any two mass values on the graph where one is half the other. To calculate $t_{\frac{1}{2}}$ from a graph always choose sensible values.

The graph may have a vertical (y) axis of radioactivity measured in cps. This makes no difference as the radioactive emissions from the isotope depend on the *mass* of isotope present.

Revision questions

NOTE: Questions 9 to 15 are for Double Award Science only, not GCSE Chemistry.

1 State **two** objections that residents in a small village might have to the siting of a blast furnace in the countryside near their village. *[2]*

2 Explain why magnesium is used in flares. *[1]*

3 Name the raw materials added to a limekiln. *[3]*

4 What hazards do the following symbols indicate?

a

b

c *[3]*

5 State the raw materials used in the extraction of aluminium from its ore. *[3]*

6 What is the main cost factor in the refining of copper? *[1]*

7 Name the raw material used in the chloralkali industry. *[1]*

8 State **one** benefit to a town of having an industrial chemical plant in the local area. *[1]*

9 Name **three** synthetic materials. *[3]*

10 Classify the following plastics as thermosetting or thermoplastic.
a PVC **b** bakelite
c Melamine **d** polythene *[4]*

11 What is a composite material? *[2]*

12 Name **two** composite materials. *[2]*

13 Which of the three types of radiation has the highest penetrating power? *[1]*

14 What type of decay does each of the following isotopes undergo?
a carbon-14 **b** uranium-238 *[2]*

15 A radioactive material has a half-life of 42 days. How long does it take for a 20 g sample of the material to decay to 2.5 g? *[2]*

Appendix 1

Colours of common elements and compounds

Metal elements

Metals are usually described as silvery grey, except copper which is a red-pink colour and iron which is dark grey.

Non-metal elements

The colours of common solids, liquids and gases are given in the following table.

Solids	Liquids	Gases
carbon = black	bromine = red/brown	nitrogen = colourless
sulphur = yellow		oxygen = colourless
iodine = grey/black		fluorine = pale yellow
		chlorine = yellow/green

Compounds

The usual colours of the compounds of metals are given in the table below. Some exceptions are listed overleaf. The colour of the solution applies only if the compound is soluble.

NOTE: Always check the back of the *Data Leaflet* provided in the exam as it gives the names of soluble and insoluble compounds.

Element	Colour of solid compounds	Colour of solution if soluble
sodium	white	colourless
potassium	white	colourless
magnesium	white	colourless
calcium	white	colourless
barium	white	colourless
aluminium	white	colourless
zinc	white	colourless
iron(II)	green	pale green
iron(III)	red-brown	yellow
copper(II)	blue	blue
silver(I)	white	colourless

Exceptions

- cobalt(II) chloride – hydrated it is pink; anhydrous it is pale blue
- copper(II) carbonate is green
- copper(II) oxide is black
- copper(II) sulphate – hydrated it is blue; anhydrous it is white
- zinc oxide changes from white to yellow on heating and back to white on cooling
- silver(I) bromide is cream
- silver(I) iodide is yellow

Appendix 2
Methods of separating mixtures

Definitions

- A mixture is defined as two or more substances mixed together that are usually easy to separate.
- Solids that dissolve in water are described as **soluble** and solids that do not dissolve in water are described as **insoluble**. A solid that dissolves is called a **solute** and the liquid in which it dissolves (usually water) is called the **solvent**. The resulting mixture is called a **solution**.
- Liquids that mix (for example, alcohol and water) are described as **miscible**. Liquids that do not mix (for example, oil and water) are described as **immiscible**.

The method of separating a mixture depends on the properties of the substances in the mixture.

Filtration

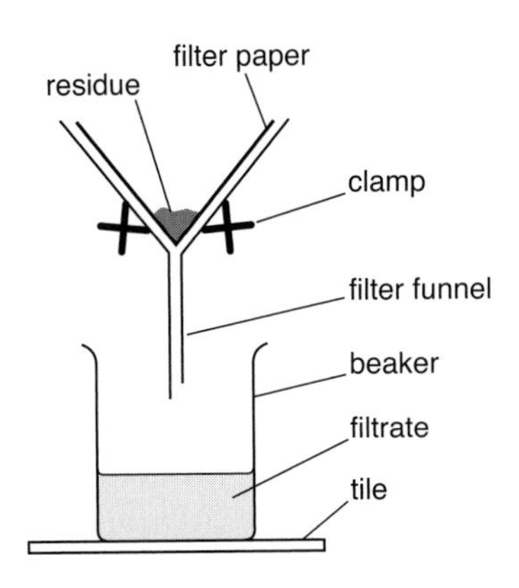

Filtration separates an insoluble solid from a liquid (suspension). It may be used to separate a mixture of soluble and insoluble solids once added to the solvent. The filtered liquid is called the **filtrate** and the solid remaining in the filter paper is called the **residue**.

Recrystallisation

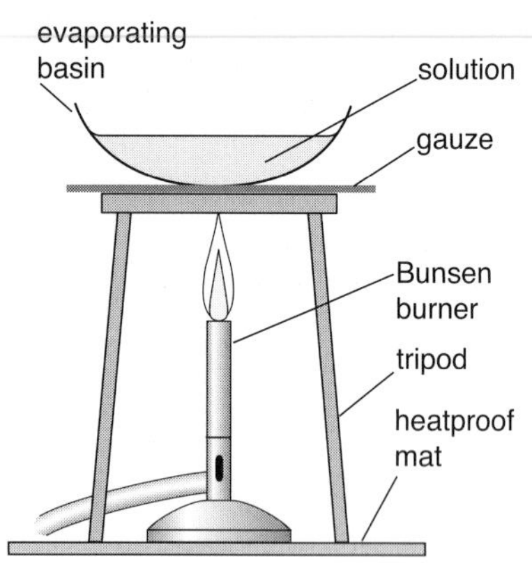

Recrystallisation retrieves the solute from a solution (the solution is usually the filtrate from a filtration).

The solution is heated gently in an evaporating basin until it is reduced to half volume, and then left aside to cool and crystallise (this ensures that the water of crystallisation is included in the crystals – heating to dryness would remove the water of crystallisation). The crystals are then filtered off and dried between two sheets of filter paper.

Separating funnel

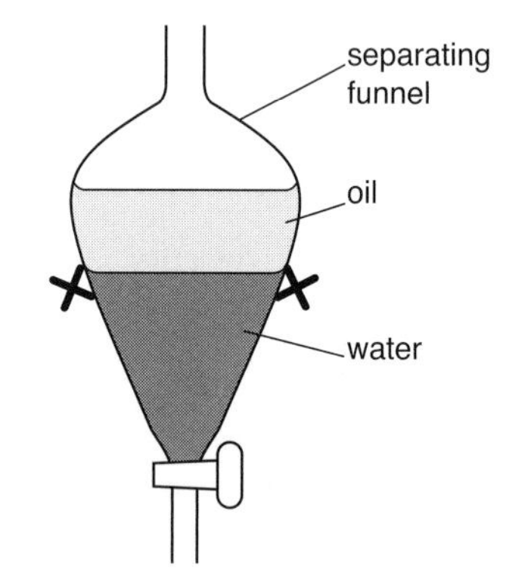

A separating funnel separates immiscible liquids, for example oil and water, based on differences in their densities, i.e. one liquid floats on top of the other.

Simple distillation

Simple distillation separates a solvent from a solution or two miscible liquids based on differences in their boiling points. (If the solvent is flammable, for example ethanol, a water bath is used to heat the mixture).

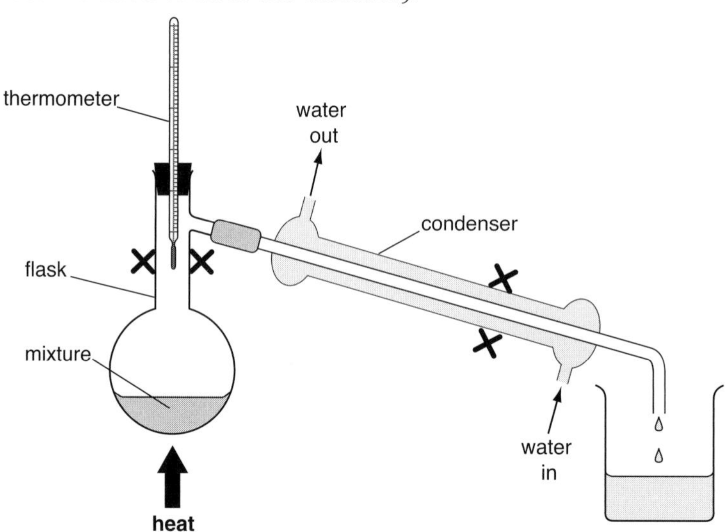

Fractional distillation

Fractional distillation separates miscible liquids, for example crude oil, or components of air based on differences in their boiling points. This uses the same apparatus as simple distillation with a fractionating column on top of the flask.

Chromatography

Chromatography separates more than one solute in a solution on the basis of differences in their solubilities in the solvent. The separation is often carried out on a paper **chromatogram**.

Answers to revision questions

1 States of matter

1 as temperature increases *[1]*, volume increases *[1]* *or* as temperature decreases *[1]*, volume decreases *[1]*

2 close packed *[1]*, regular arrangement *[1]*

3 spaces between particles *[1]*, particles can be pushed closer together *[1]*

4 *two from:* fixed volume, cannot be compressed, fluid, can flow, not fixed shape, takes shape of bottom and sides of container *[2]*

5 $\frac{P_1V_1}{T_1} = \frac{P_2V_2}{T_2}$ *[1]*

$\frac{2.5 \times 150}{200} = \frac{1.5 \times V_2}{150}$ *[1]*

$V_2 = \frac{2.5 \times 150 \times 150}{200 \times 1.5}$ *[1]*

$V_2 = 187.5\ \text{cm}^3$ *[1]*

6 melting: the change from a solid to a liquid *[1]*, subliming: the change from a solid to a gas *[1]*

7 *one from:* dry ice, carbon dioxide, iodine *[1]*

8 $\frac{P_1V_1}{T_1} = \frac{P_2V_2}{T_2}$ *[1]*

$\frac{2.5 \times 25}{200} = \frac{2 \times V_2}{300}$ *[1]*

$V_2 = \frac{2 \times 25 \times 300}{200 \times 2}$ *[1]*

$V_2 = 37.5\ \text{dm}^3$ *[1]*

9 condensing *[1]*

10 a a mark on the tube closer to the end soaked in concentrated hydrochloric acid *[1]*

b HCl particles are larger/heavier *[1]* and move more slowly *[1]* *or* NH_3 particles are smaller/lighter *[1]* and move more quickly *[1]*

11 Brownian motion *[1]*

12 water particles bombard *[1]* sugar particles, sugar particles break up *[1]* and move between the water particles *[1]*

13 a solid *[1]* **b** liquid *[1]* **c** gas *[1]*

14 dry ice *[1]*

15 grey *[1]* solid *[1]* changes to purple *[1]* gas *[1]*

2 The Periodic Table

1 a alkali metals *[1]*

b alkaline earth metals *[1]*

c halogens *[1]*

d Noble gases *[1]*

2 Mg *or* Na *[1]*

3 phosphorus, P *[1]*

4 John Newlands *[1]*

5 a pale yellow gas *[1]*

b yellow-green gas *[1]*

c red-brown liquid *[1]*

d grey solid *[1]*

e colourless gas *[1]*

6 a Rb^+ *[1]* **b** Sr^{2+} *[1]*

c Li^+ *[1]* **d** F^- *[1]*

7 germanium *[1]*; exhibits properties of both metals and non-metals *[1]*

8 goes dull *[1]*

9 atomic size decreases *[1]*

10 a hydrogen *[1]*, oxygen *[1]*, nitrogen *[1]*

b carbon *or* sulphur *[1]*

c aluminium *[1]*

d sulphur, SO_2 *or* carbon, CO_2 *or* nitrogen, NO_2 *or* silicon, SiO_2 *[1]*

e magnesium *or* aluminium *[1]*

11 an oxide which reacts with an acid *[1]*

12 atomic number *[1]*

13 reacts with both acids and alkalis *[1]*

14 increases up to Group IV *[1]*, then decreases again to 1 for Group VII *[1]*

15 fluorine *[1]*

3 Atomic structure and bonding

1 number of protons *[1]*

2 **a** relative mass = 1; relative charge = +1 *[1]*
b relative mass = $\frac{1}{1840}$; relative charge = −1 *[1]*
c relative mass = 1; relative charge = 0 *[1]*

3 atoms of the same element with the same number of protons but different numbers of neutrons *[1]*

4 (diagrams to match electronic configurations)
a 2, 8, 5 *[1]* **b** 2, 1 *[1]*
c 2, 6 *[1]* **d** 2, 8, 8, 1 *[1]*
e 2, 8, 8 *[1]* **f** 2 *[1]*
g 2, 8, 3 *[1]* **h** 2, 8, 1 *[1]*

5 **a** 2, 8 *[1]* **b** 2, 8 *[1]*
c 2, 8 *[1]* **d** 2, 8 *[1]*
e 2, 8, 8 *[1]*

6 when molten the ions *[1]* can move *[1]* and carry charge *[1]*

7 sharing *[1]* of a pair *[1]* of electrons *[1]*

8 **a** 7 electrons in outer shell of each Cl *[1]*
2 electrons shared in bond *[1]*

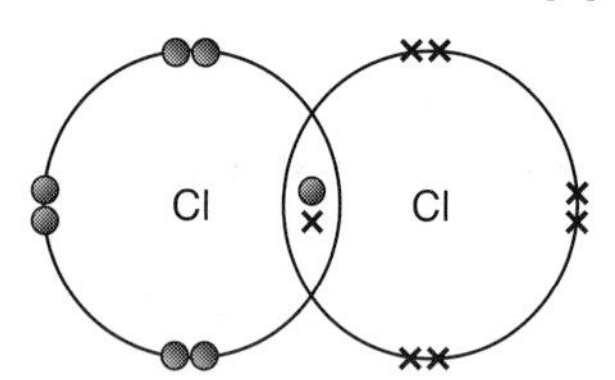

b 6 electrons in outer shell of O *[1]*
1 electron in outer shell of H *[1]*
2 H and 1 O atom *[1]*
2 electrons shared in each bond *[1]*

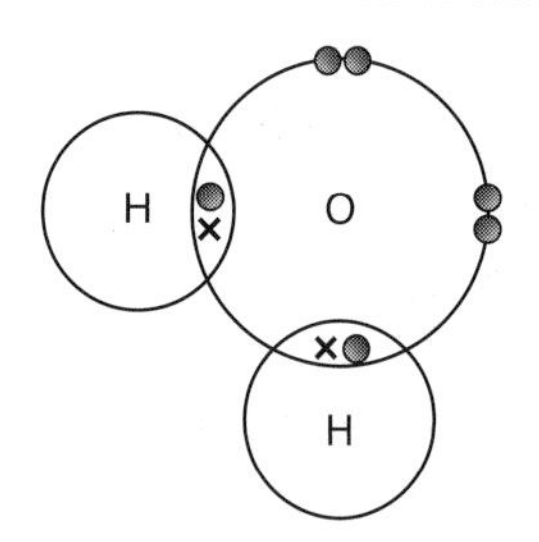

9 little energy required *[1]* to break the weak bonds *[1]* between chlorine molecules *[1]*

10 each carbon atom *[1]* covalently *[1]* bonded to four *[1]* other carbon atoms in a tetrahedral structure *[1]*

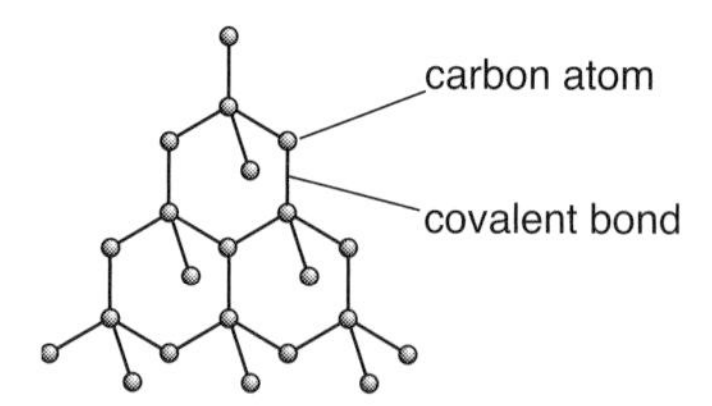

11 delocalised electrons *[1]* can move *[1]* and carry charge *[1]*

12 can be hammered into shape *[1]*

13 layers *[1]* of positive centres/ions *[1]* held together by a sea of delocalised electrons *[1]*

14 20 protons and 20 neutrons *[1]* in the nucleus *[1]*
20 electrons *[1]* arranged 2, 8, 8, 2 *[1]*

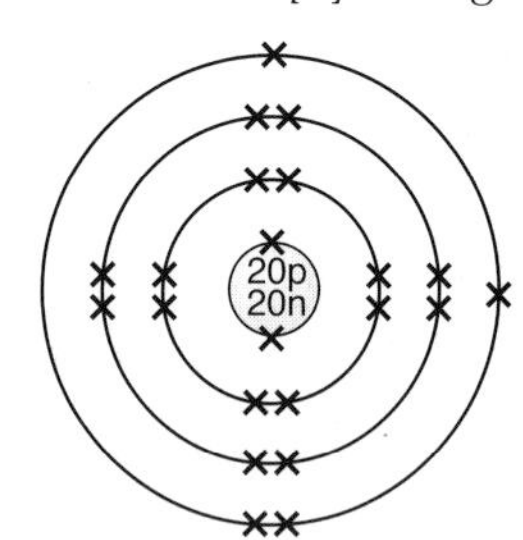

15 See table below.

Particle	Atomic number	Mass number	Number of protons	Number of neutrons	Number of electrons	Electronic configuration
X	**11**	**23**	11	12	**10**	2, 8
Y	17	**35**	**17**	18	17	**2, 8, 7**
Z	**8**	16	8	**8**	10	**2, 8**

4 Formulae and equations

1 a NaF *[1]* **b** MgO *[1]*
c K_2O *[1]* **d** $BaCl_2$ *[1]*

2 a $CuCl_2$ *[1]* **b** ZnO *[1]*
c $CuSO_4$ *[1]* **d** $Fe(OH)_3$ *[1]*

3 sodium carbonate: Na_2CO_3 *[1]*; sodium hydrogen carbonate: $NaHCO_3$ *[1]*

4 a carbon dioxide *[1]*
b potassium nitrate *[1]*
c copper carbonate *[1]*
d hydrogen fluoride *[1]*
e magnesium sulphate *[1]*

5 a $(NH_4)_2SO_4$ *[1]* **b** SO_2 *[1]*
c $Ca(HCO_3)_2$ *[1]* **d** $Al_2(SO_4)_3$ *[1]*

6 $S_2O_3^{2-}$ *[1]*

7 a hydroxide *[1]* **b** oxide *[1]*
c chloride *[1]* **d** aluminium *[1]*
e sulphate *[1]*

8 a $2KOH + H_2SO_4 \rightarrow K_2SO_4 + 2H_2O$ *[3]*
b $2Ca + O_2 \rightarrow 2CaO$ *[3]*
c $2Al + 3Cl_2 \rightarrow 2AlCl_3$ *[3]*

9 $CuCO_3 \rightarrow CuO + CO_2$ *[2]*

10 $2C_2H_6 + 7O_2 \rightarrow 4CO_2 + 6H_2O$
(*or* $C_2H_6 + 3\frac{1}{2}O_2 \rightarrow 2CO_2 + 3H_2O$) *[3]*

11 $Na_2S_2O_3 + 2HCl \rightarrow 2NaCl + S + SO_2 + H_2O$ *[3]*

12 a $Ca(OH)_2 + 2HCl \rightarrow CaCl_2 + 2H_2O$ *[3]*
b $Al_2O_3 + 3H_2SO_4 \rightarrow Al_2(SO_4)_3 + 3H_2O$ *[3]*
c $Zn + 2HCl \rightarrow ZnCl_2 + H_2$ *[3]*

13 $N_2 + 3H_2 \rightarrow 2NH_3$ *[3]*

14 $Cu^{2+} + 2OH^- \rightarrow Cu(OH)_2$ *[3]*

15 a $Mg + Cu^{2+} \rightarrow Mg^{2+} + Cu$ *[3]*
b $Zn + 2H^+ \rightarrow Zn^{2+} + H_2$ *[3]*
c $CO_3^{2-} + 2H^+ \rightarrow CO_2 + H_2O$ *[3]*

5 Acids, bases and salts

1 a soluble *[1]* base *[1]*

2 *two from:* copper oxide, copper carbonate, copper hydroxide *[2]*

3 $Al_2O_3 + 2KOH \rightarrow 2KAlO_2 + H_2O$ *[3]*

4 yellow *[1]* precipitate *[1]*

5 H^+ *[1]*

6 *any two from:* sulphur dioxide, nitrogen dioxide, carbon dioxide, silicon dioxide *[2]*

7 barium chloride *or* barium nitrate *[1]* with any soluble sulphate *[1]*, for example sodium sulphate, magnesium sulphate etc.

8 $Ag^+ + Cl^- \rightarrow AgCl$
[1] for Ag^+ and Cl^-, *[1]* for AgCl

9 a colourless *[1]* **b** colourless *[1]*
c pink *[1]*

10 a red *[1]* **b** orange *[1]*
c yellow *[1]*

11 a $Mg + 2HCl \rightarrow MgCl_2 + H_2$ *[3]*
b $Mg(OH)_2 + 2HCl \rightarrow MgCl_2 + 2H_2O$ *[3]*
c $Na_2CO_3 + 2HCl \rightarrow 2NaCl + CO_2 + H_2O$ *[3]*

12 potassium sulphate *[1]*

13 hydrogen *[1]*

14 a pH 3–5 *[1]* **b** pH 7 *[1]*
c pH 12–14 *[1]*

15 a residue *[1]* **b** filtrate *[1]*
c filter *[1]* to remove crystals and dry between two sheets of filter paper or in a desiccator or low-temperature oven *[1]*

6 Electrolysis

1 red-brown *[1]* pungent *[1]* gas *[1]*

2 $2Cl^- \rightarrow Cl_2 + 2e^-$
[1] for Cl^-, *[1]* for $Cl_2 + e^-$, *[1]* for balancing

3 decomposition *[1]* using a direct current of electricity *[1]*

4 graphite *[1]*

5 a positive electrode *[1]*
b negative electrode *[1]*

6 liquid *[1]* which conducts electricity *[1]* and is decomposed *[1]* by it

7 anode: impure *[1]* copper *[1]*
cathode: pure *[1]* copper *[1]*

8 conducts electricity *[1]*; inert *or* unreactive *[1]*

9 bauxite *[1]*

10 cryolite *[1]*

11 $2H^+ + 2e^- \rightarrow H_2$
[1] for $H^+ + e^-$, *[1]* for H_2, *[1]* for balancing

12 $C + O_2 \rightarrow CO_2$ *[2]* *or* graphite *[1]* anode burns *[1]* away

13 $2Br^- \rightarrow Br_2 + 2e^-$ *[3]*

14 anode: oxygen *[1]*; cathode: hydrogen *[1]*

15 $4OH^- \rightarrow O_2 + 2H_2O + 4e^-$
[1] for OH^-, *[1]* for $O_2 + H_2O + e^-$, *[1]* for balancing

7 Water

1 shiny *[1]* surface *[1]* becomes dull *[1]*

2 water which does not lather readily *[1]* with soap *[1]*

3 calcium hydrogen carbonate *[1]*

4 hydrated *[1]* sodium carbonate *[1]*

5 decreases *[1]*

6 **a** liquid *[1]* in which substances dissolve *[1]*
b solid *[1]* that dissolves *[1]*
c mixture *[1]* of solute dissolved in a solvent *[1]*
d solution that cannot dissolve *[1]* any more solute *[1]*

7 water chemically bonded *[1]* in the crystal structure *[1]*

8 **a** blue *[1]* **b** pink *[1]*

9 heat *[1]* or concentrated sulphuric acid *[1]*

10 sodium carbonate *[1]*;
$2NaOH + CO_2 \rightarrow Na_2CO_3 + H_2O$ *[3]*

11 g/100 g water *[1]*

12 eutrophication *[1]*

13 **a** limestone *[1]* reacts *[1]* with rainwater *[1]* containing carbon dioxide *[1]* to form calcium hydrogen carbonate *[1]*
b carbonate ions, CO_3^{2-} *[1]* from washing soda *[1]* react *[1]* with calcium ions, Ca^{2+} *[1]* from hard water *[1]* to form insoluble *[1]* calcium carbonate, $CaCO_3$ *[1]* *[max. 4]*

14 **a** sample A *[1]* **b** sample C *[1]*
c sample B *[1]*
d all hardness removed *[1]* by boiling *[1]*

15 3 g (40.5 g/100 g water at 45 °C and 34.5 g/100 g water at 25 °C. The difference is 6 g/100 g water, so 3 g per 50 g)

8 Quantitative chemistry

1 carbon *[1]* 12 *[1]*

2 equal volumes of gases *[1]* under the same conditions of temperature and pressure *[1]* contain the same number of particles *[1]*

3 **a** 98 *[1]* **b** 74 *[1]* **c** 342 *[1]*
d 138 *[1]* **e** 162.5 *[1]*

4 RFM $CaCO_3$ = 100 *[1]*, moles $CaCO_3$ = 0.05 *[1]*, moles CaO = 0.05 *[1]*, RFM CaO = 56 *[1]*, mass CaO = 2.8 g *[1]*

5 RFM Mg = 24 *[1]*, moles Mg = 0.05 *[1]*, moles MgO = 0.05 *[1]*, RFM MgO = 40 *[1]*, mass MgO = 2 g *[1]*

6 RFM Al = 27 *[1]*, moles Al = 2000 *[1]*, moles of Fe_2O_3 = 1000 *[1]*, RFM Fe_2O_3 = 160 *[1]*, mass of Fe_2O_3 = 160 kg *[1]*

7 moles of Al = 2000 *[1]*, moles of Fe = 2000 *[1]*, mass of Fe = 2000 × 56 = 112 kg *[1]*

8 14 × 2 *[1]* = 28 cm^3 *[1]*

9 nucleus *[1]*

10 **a** 72.4% *[1]*
b moles of Fe = 1.293 *[1]*, moles of O = 1.725 *[1]*, simplest ratio = 3:4 *or* Fe_3O_4 *[1]*

11 **a** RFM Mg = 24 *[1]*, moles Mg = 0.025 *[1]*, moles HCl = 0.05 *[1]*, volume of HCl = 500 cm^3 *[1]*
b moles H_2 = 0.025 *[1]*, volume of H_2 = 600 cm^3 *or* 0.6 dm^3 *[1]*

12 **a** 0.004875 *[2]* **b** 0.00975 *[2]*
c 0.39 mol/dm^3 *[2]*

13 **a** 0.0035 *[2]* **b** 0.0035 *[2]*
c 0.014 *[2]* **d** 40 *[2]*
e 23 *[2]* **f** sodium *or* Na *[1]*

14 rinse with deionised water *[1]*; rinse with solution *[1]*; fill with solution *[1]* ensuring jet is filled *[1]* no air bubbles *[1]* *[max. 3]*

15 **a** 0.212 g *[2]* **b** 0.002 *[2]* **c** 0.288 g *[2]*
d 0.016 *[2]* **e** 8 *[2]*

9 Rates of reaction

1 manganese dioxide *or* manganese(IV) oxide *[1]*

2 sulphur *[1]*

3 gas syringe *[1]*

4 iron *[1]*

5 rate increases *[1]*

6 $Ag^+ + e^- \rightarrow Ag$
[1] for $Ag^+ + e^-$, [1] for Ag

7 hydrogen *[1]*

8 *one from:* size of solid particles, presence of a catalyst, light *[1]*

9 loss of mass *[1]* using a conical flask on a balance *[1]* *or* gas volume *[1]* using a gas syringe *[1]*

10 **a** to give a higher rate of reaction *[2]*
b higher pressure is too expensive *[2]*

11 80 s *[1]*

12 80 cm^3 *[1]*

13 starts at 0 *[1]*, gas volume higher *[1]*, levels off earlier *[1]*, at same gas volume *[1]*

14 starts at 0 *[1]*, gas volume lower *[1]*, levels off later *[1]*, at same gas volume *[1]*

15 150 cm^3 *[2]*

10 Metals and their compounds

1 can be drawn into wires *[1]*

2 *four from:* floats/moves about the surface, fizzes, lilac flame, eventually disappears, explodes, heat released *[4]*

3 $2K + 2H_2O \rightarrow 2KOH + H_2$ *[1]*

4 *three from:* solid appears, blue solution, fades to green or colourless, heat released *[3]*

5 *two from:* Ca^{2+}, Al^{3+}, Zn^{2+} *[2]*

6 coke or carbon *[1]*, iron ore or haematite *[1]*, limestone or calcium carbonate *[1]*

7 $Fe^{2+} + 2OH^- \rightarrow Fe(OH)_2$ *[3]*

8 copper carbonate: green *[1]*, copper oxide: black *[1]*

9 $CaCO_3 \rightarrow CaO + CO_2$ *[2]*

10 *two from:* expands, hisses, crumbles, heat released *[2]*

11 limewater *[1]*

12 hydrated *[1]* iron(III) oxide *[1]*

13 magnesium reacts before the iron *[1]*, sacrificial protection *[1]*

14 green *[1]* solution changes to yellow *[1]*

15 **a** flame test rod *[1]* dipped in deionised water *[1]*, into the flame *[1]*, dip rod into sample *[1]* into flame and observe colour *[1]*
b copper(II) chloride *[1]*

11 Non-metals

1 **a** manganese dioxide and hydrogen peroxide *[2]*
b calcium carbonate and hydrochloric acid *[2]*
c zinc and hydrochloric acid *[2]*
d sodium chloride and concentrated sulphuric acid *[2]*
e ammonium chloride and sodium hydroxide *[2]*
f potassium permanganate and concentrated hydrochloric acid *[2]*

2 **a** lighted splint *[1]* 'pops' *[1]*
b limewater *[1]* changes from colourless *[1]* to milky *[1]*
c glowing splint *[1]* re-lights *[1]*
d glass rod *[1]* dipped in concentrated *[1]* ammonia *[1]* held near gas produces white *[1]* smoke *[1]* *maximum [4] out of [5]*

3 $NH_3 + H_2O \rightarrow NH_4OH$ *[2]* *or* hydroxide ions *[1]* in water *or* aqueous *[1]*

4 *two from:* fire extinguishers, carbonated drinks, dry ice *[2]*

5 *any four from:* blue *[1]* solution, blue *[1]* precipitate *[1]*, re-dissolves *[1]* to form a dark blue *[1]* solution

6 **a** ammonia, air *[2]* **b** platinum/rhodium *[1]*
c 900–1000 °C *[1]*

7 Contact process *[1]*

8 *any three from:* gloves, safety glasses, lab. coat *[1]*; add acid to water *[1]* slowly/dropwise *[1]* with stirring *[1]*

9 hydrochloric acid *[1]*

10

Halogen	State at room temperature and pressure	Colour	
chlorine	gas	yellow-green	[2]
fluorine	gas	pale yellow	[2]
bromine	liquid	red-brown	[2]
iodine	solid	dark grey	[2]

11 **a** $2Mg + CO_2 \rightarrow 2MgO + C$ *[3]*
b bright white light *[1]*, white solid *[1]*, black specks *[1]*

12 $N_2 + 3H_2 \rightarrow 2NH_3$ *[3]*

13 *three from:* bubbles of gas, heat released, solid disappears, colourless solution *[3]*

14 different forms of the same element *[1]* in the same physical state *[1]*

15 rhombic *[1]*, monoclinic *[1]*, plastic *[1]*

12 Organic chemistry

1 **a** C_2H_6 *[1]* **b** C_2H_4 *[1]*
c C_4H_{10} *[1]* **d** CH_4 *[1]*

2 red-brown *[1]* solution changes to colourless *[1]*

3 compounds containing only *[1]* carbon and hydrogen *[1]*

4 **a** propane *[1]* **b** ethanol *[1]*
c ethene *[1]* **d** ethyl ethanoate *[1]*

5 $C_2H_5OH + 3O_2 \rightarrow 2CO_2 + 3H_2O$ *[3]*

6 ethanol *[1]*

7 fractional *[1]* distillation *[1]*

8 *any three from:* decomposition *[1]* by heat *[1]* of longer chain hydrocarbons *[1]* into shorter *[1]* more useful *[1]* ones

9 $2CH_3COOH + Mg \rightarrow (CH_3COO)_2Mg + H_2$ *[3]*

10 family of organic compounds with the same general formula *[1]* which differ by a CH_2 unit *[1]* and show a gradation in their physical properties *[1]* and similar chemical properties *[1]*

11 **a** polythene *[1]*
b polyvinyl chloride (PVC) *[1]*

12 global warming *or* climate change *[1]*, increased sea levels *[1]*, melting of polar ice caps *[1]*

13 carbon monoxide *[1]*, water *[1]*

14 fuel formed from dead plants and animals *[1]* over millions of years *[1]* under the action of pressure and heat *[1]*

15 aircraft fuel *[1]*

13 Chemical change

1 **a** gives out *[1]* heat *[1]*
b takes in *[1]* heat *[1]*

2 exothermic: **b**, **d** *[2]*; endothermic: **a**, **c** *[2]*

3 carbon dioxide *[1]*

4 *one from:* magnesium, zinc *[1]*

5 $S + O_2 \rightarrow SO_2$ *[2]*; $SO_2 + H_2O \rightarrow H_2SO_3$ *[2]*

6 **a** fuel reacting with oxygen *[1]* releasing heat *[1]* forming oxides *[1]*
b iron reacting with oxygen and water *[1]* forming hydrated iron(III) oxide *[1]*
c oxidation and reduction occurring in the same reaction *[1]*

7 $Al^{3+} + 3e^- \rightarrow Al$ *[3]*

8 $CuCO_3 \rightarrow CuO + CO_2$ *[2]*; green solid *[1]* changes to black *[1]*

9 **a** $Zn \rightarrow Zn^{2+} + 2e^-$ *[2]*
b $Cu^{2+} + 2e^- \rightarrow Cu$ *[2]*
c $Zn \rightarrow Zn^{2+} + 2e^-$ is oxidation (loss of electrons) *[1]*

10 *one from:* sodium carbonate, potassium carbonate *[1]*

11 defoliates trees (loss of leaves) *[1]*, destroys limestone buildings and statues *[1]*, kills fish in lakes and rivers *[1]*

12 nitrogen gains hydrogen *[1]*, gain of hydrogen is reduction *[1]*

13 *any three from:* yellow solid *[1]* burns with a blue *[1]* flame releasing heat *[1]* producing a colourless/misty *[1]* pungent *[1]* gas *[1]*

14 sulphur dioxide produced in one country *[1]* can cause acid rain in another country *[1]*

15 water *[1]*, air *or* oxygen *[1]*

14 Rocks, materials and radioactivity

1 *three from:* nylon, polythene, iron, aluminium, PVC, polystyrene *[3]*

2 *two from:* increased heavy traffic, noise pollution, destroys hedgerows, destruction of habitats, eyesore *[2]*

3 burns with a bright white flame *[1]*

4 limestone *[1]*, coke *[1]*, hot air *[1]*

5 thermosetting: **b**, **c** *[2]*; thermoplastic: **a**, **d** *[2]*

6 material which combines the properties of more than one material *[1]* to produce a more useful material for a particular use *[1]*

7 **a** toxic *[1]* **b** explosive *[1]*
c harmful *[1]*

8 gamma *[1]*

9 **a** beta *[1]* **b** alpha *[1]*

10 3 half-lives *[1]* $= 3 \times 42 = 126$ days *[1]*

11 *two from:* glass fibre, reinforced concrete, reinforced glass, bone *[2]*

12 alumina *[1]*, cryolite *[1]*, graphite *[1]*

13 electricity *[1]*

14 seawater *or* sodium chloride solution *[1]*

15 higher employment *[1]*

Index